Hydrogen Fuel Used in Car

Hydrogen Fuel Used in Car

Dr. Seema Tiwari
Mr. John Abbas (IBIS Group)
Prof. (Dr.) B. K. Sarkar (Patent Guru/ Expert)
Naveen Rana

NAMYA PRESS.com

Published by

Namya Press

India: 213, Vardaan House, 7/28 Ansari Road, Daryaganj, Delhi-110002
Email: namyapress@gmail.com
Website: https://namyapress.com

Edition	:	First Published in 2023 (Namya Press)
Title	:	Hydrogen Fuel Used in Car
Editors	:	Dr. Seema Tiwari, Mr. John Abbas Prof. (Dr.) B. K. Sarkar, Naveen Rana
ISBN	:	978-93-5545-118-7
Copyright	:	© Dr. Seema Tiwari, Mr. John Abbas Prof. (Dr.) B. K. Sarkar, Naveen Rana 2023 All Rights Reserved

Preface

It's with great happiness that I would like to acknowledge a great deal of people that get helped me extremely through the entire difficult, challenging, but a rewarding and interesting path towards some sort of Edited Book Without having their help and support, none of this work could have been possible.

Contents

1

Hydrogen Fuel Cell Vehicles: Current Status and Future Prospect

Yogesh Manoharan

Combustion and Sustainable Energy Laboratory (ComSEL), Department of Mechanical Engineering, Arkansas Tech University, 1811 N Boulder Ave, Russellville, AR 72801, USA; ymanoharan@atu.edu (Y.M.);

Himanshu Kandpal

Quantum University, Dehradun Highway, Mandawar, Roorkee, Uttarakhand 247167, India

Dr. Seema Tiwari

Associate Professor Applied Science & General Engineering Army Institute of Technology, Pune.

Mr. John Abbas

IBIS Group, Lucknow, UP, India.

Prof. (Dr.) B. K. Sarkar (Patent Guru/ Expert)

MES, Geh Research, India, USA, Singapore, Japan, Austraila, Germany.

Prof. (Dr.) Reena Singh,

MES's, Pillai HOC College of Engineering & Technology, Rasayani Taluka Panvel, Dist, Navi Mumbai, Maharashtra 410207, India

ABSTRACT

The hazardous effects of pollutants from conventional fuel vehicles have caused the scientific world to move towards environmentally friendly energy sources. Though we have various renewable energy sources, the perfect one to use as an energy source for vehicles is hydrogen. Like electricity, hydrogen is an energy carrier that has the ability to deliver incredible amounts of energy. Onboard hydrogen storage in vehicles is an important factor that should be considered when designing fuel cell vehicles. In this study, a recent development in hydrogen fuel cell engines is reviewed to scrutinize the feasibility of using hydrogen as a major fuel in transportation systems. A fuel cell is an electrochemical device that can produce electricity by allowing chemical gases and oxidants as reactants. With anodes and electrolytes, the fuel cell splits the cation and the anion in the reactant to produce electricity. Fuel cells use reactants, which are not harmful to the environment and produce water as a product of the chemical reaction. As hydrogen is one of the most efficient energy carriers, the fuel cell can produce direct current (DC) power to run the electric car. By integrating a hydrogen fuel cell with batteries and the control system with strategies, one can produce a sustainable hybrid car.

Keywords: *hydrogen; fuel cell; battery; clean transportation; electric vehicle*

INTRODUCTION

Energy is a factor determining a country's economy, infrastructure, transportation, and standard of life. The problem faced globally is the disparity between the consumption and the availability of energy. All nations are presently depending upon fossil fuel for the energy production, and these fossil fuels are not sustainable sources. To supply the energy demands of the more rapidly increasing global population, it is essential to upgrade to an alternative, sustainable energy source that does not negatively affect the environment [1,2]. In recent decades, the United States has been putting emphasis on the environmental impact of the transportation sector and reducing petroleum dependence. Depletion on non-renewable conventional fuels is one of the main issues of the modern energy scenario, which makes the state of the energy industry unsustainable, and it also causes environmental problems such as the greenhouse effect [3,4]. Today, the proportion of fossil fuel use is still high, and it is projected that it will account for approximately 75% of energy production in 2050 [5]. Overall, the current energy scenario has many downsides. However, there exist many sustainable energy sources, and if their use increases, the scenario will be much more optimistic for future generations. It is predicted by environmentalists that the worst-case scenario of global warming and its effect will not be reached because of different initiatives. Over the last two decades, vehicles have become more fuel efficient, and hybrid electric vehicles are becoming more common. One of the fastest growing alternative energies in vehicles is electricity. As with conventional energy sources such as petroleum and coal, electricity is not a primary energy source. A fully charged battery is an energy carrier. Battery electric vehicles (BEVs) are highly efficient at converting energy from the grid into tractive force, and they can recover energy during drives by utilizing regenerative braking. One major drawback of BEVs is that they usually have a limited range due to the size and the cost of batteries necessary for vehicle power and energy requirements. The "refueling" of the battery systems can also take several hours, rather than a few minutes with a conventional vehicle (CV). To use advantages of both electric and conventional vehicles and to bridge the gap between CVs and BEVs, an alternative is considered. Hydrogen is a chemical energy carrier that has the capability to produce electricity up to 39.39 kWh/kg, which surpasses the energy density of most batteries. A fuel cell (FC) has a direct analogy to an internal combustion engine (ICE). An ICE converts chemical energy stored in the fuel supplied to the engine to produce rotational mechanical energy [6]. The rotational energy produced is then either used to propel a vehicle or focused through a generator and converted into electrical energy. An FC acts much in the same way as an ICE in that chemical energy is directly converted into electrical energy in the FC, but in an environmentally friendly process [7–12]. Unlike a battery that drains while it is used to power electrical components, internal combustion engines and fuel cells act as continually operational power sources as long as fuel is being provided to them [13]. Hence, it is projected that the hydrogen fuel cell can overcome the disadvantages of BEVs, making hydrogen the transportation fuel of the future.

HYDROGEN AS A TRANSPORTATION FUEL

Hydrogen is the simplest form of all molecules; it has the lowest energy content by volume, but it has the highest energy content of any fuel by weight. It is available in the atmosphere as gas and in water as liquid. Due to the high energy content of hydrogen, it is employed as a fuel in applications such as FCs and rockets. Hydrogen creates zero harmful emissions, which is one of the most significant drawbacks of fossil fuels, and the heating value of hydrogen is three times higher than that of petroleum. There is a hefty production cost involved with hydrogen, since it is a manmade fuel, costing about three times more than petroleum refining [14]. Substantial amounts of research are dedicated to creating an efficient and sustainable way to produce hydrogen and applications for hydrogen in transportation engines. Automobile manufacturers such as Honda, Toyota, and Hyundai have started to manufacture fuel cell vehicles (FCVs) with hydrogen as fuel. These FCVs are currently available in North America, Asia, and Europe, and have primarily been bought by early adopters. The current consumers—early adopters—are primarily highly educated people, high income families, those with larger household units, those willing to change their lifestyle, and those with other similar attributes [15]. As of June 2018, there have been over 6500 FCVs sold to consumers. California is the leading market for FCVs, with nearly 3000 FCVs being delivered there out of the 5233 vehicles sold globally due to the state housing the largest network of hydrogen refueling stations in the world and automakers selling the vehicles there. Presently, several automakers are promoting FCVs to consumers, which are often compared to BEVs. Both BEVs and FCVs offer zero tailpipe emissions, the ability to be fueled using renewable and sustainable energy sources, and use electric motors. The most notable differences between FCVs and BEVs are the driving range and the refueling style. FCVs have above a 300-mile driving range and can be refueled in less than 10 min at a hydrogen refueling station, which is more comparable to a traditional ICE fossil fuel powered vehicle. Hydrogen has greater potential for use as a fuel in the future. It is estimated that, by the year 2030, the cost of fuel cells will be competitive with ICEs based on the technological improvements being made and the increased availability [14]. One of the main hurdles that mass hydrogen use faces is more efficient storage. Because of the low density of hydrogen, it cannot be stored as easily as traditional fossil fuels. Hydrogen requires compression, cooling, or a combination of them. The most favorable method of hydrogen storage is physical containment, specifically in compressed tanks, because they are readily available. All composites (Type IV) are primarily used, and sometimes metal lined composites (Type III) are used. The fill time of these tanks is competitive with fossil fuels when the hydrogen is pre-cooled. Cost is the main setback for the wide scale use of compressed hydrogen (CH_2) tanks, because the material and the assembly are expensive. Another potential setback is the public's concern of using such high pressure (70 MPa) storage tanks in vehicles [16]. An alternative to traditional CH_2 tanks that is still being researched is a tank with an internal skeleton, which is a complex design of struts in tension within the tank to resist the forces of the compressed gas. Liquid hydrogen (LH_2) storage has improved significantly, achieving the best specific mass (15%) of any other automotive hydrogen storage system [17]. Energy efficiency is decreased when liquid hydrogen is used. Boil off is an area that needs improvement before LH_2 systems can be widely accepted. A promising alternative design is a cryo-compressed tank in which hydrogen is highly compressed at cryogenic temperatures. More studies must be done on this method to determine long term durability and public acceptance of the system. Hydride storage systems require substantial research and progress to meet the requirements for large scale use. The most well studied hydride is $NaAlH_4$, but it does not have the capacity necessary for application. Based on the results of the few existing studies, it is indicated that tanks with no internal heat transfer elements could be constructed based on the moderate heat of absorption of hydrogen on surfaces [18]. Though hydrogen presence is abundant in the atmosphere, it is not in the purest form. Hydrogen can be extracted from water, hydrocarbon fuel, hydrogen sulfide, and other chemical elements [19]. The energy that is used to produce hydrogen from its associated elements requires external energy, such as thermal, electrical, photonic, and biochemical energy. The chemical element, ammonia, has a high percentage of hydrogen within and was proposed as a fuel for internal combustion engines through onboard decomposition to hydrogen and nitrogen [20,21]. Hydrogen can be extracted from either non-renewable or renewable energy sources. Hydrogen production from renewables is always environmentally friendly, whereas the hydrogen produced from non-renewables emits greenhouse gases (GHG) [22,23]. Hydrogen production from the waste biomass using electrochemical reactions is GHG-less, and it does not require a large amount of energy or high production costs. Possible biomass feedstocks are bread residue, cypress sawdust, and rice chaff [24,25]. Similar to these, newspaper could be used to produce hydrogen by direct electrolysis. Newspaper is composed of 69.2% cellulose and 11.8% lignin, and it is decomposed into monosaccharides and disaccharides as well as aliphatic keto acid in the solvent H_3PO_4 under conditions similar to electrolysis [26,27]. Hydrogen can also be generated by the electrolysis of humidified methane [28].

HYDROGEN STORAGE IN FCVs

Hydrogen storage is the one of the most important research issues in the development of FCVs. Hydrogen storage systems are under development to introduce new methods to meet the needs of customers. Due to hydrogen's low energy-density, it is difficult to store enough on-board a vehicle to obtain adequate driving-range without the storage container being too large or too heavy. Figure 1 shows a hydrogen FCV with on-board storage of compressed gas [29]. Hydrogen storage techniques and the required research in this area are summarized in the following points.

Pressurized Tank Storage

Pressurized tanks of enough strength—involving impact resistance for safety in collisions—are made of carbon-fiber wrapped cylinders. Compressed hydrogen in such tanks has been illustrated at a pressure of 34 MPa with a mass of 32.5 kg and a volume of 186 L, which is adequate for a 500-km range. The tank volume is about 90% of a 55-gallon drum, which is large for individual automobiles. While the 6 wt% goal can be obtained, the tank volume is problematic. Pressures of 70 MPa have been achieved, and in 2002 in Germany, Quantum Technology's

10,000 psi (68 MPa) on-board storage tank was certified [30]. The office of technology policy report [31] says that Toyota and Honda vehicles available for lease in late-2002 use hydrogen stored in high-pressure containers [32]. Nevertheless, unlike other gases, enough hydrogen cannot be stored in the tank because of its low density. Liquid hydrogen storage at low temperatures is not appropriate for normal vehicle use, although research on this feasibility is being continued at a low level by some manufacturers. Moreover, a liquid hydrogen storage system loses up to 1% of its storage volume per day by boiling, and liquid hydrogen requires high refrigeration to keep the hydrogen at 20 K [31].

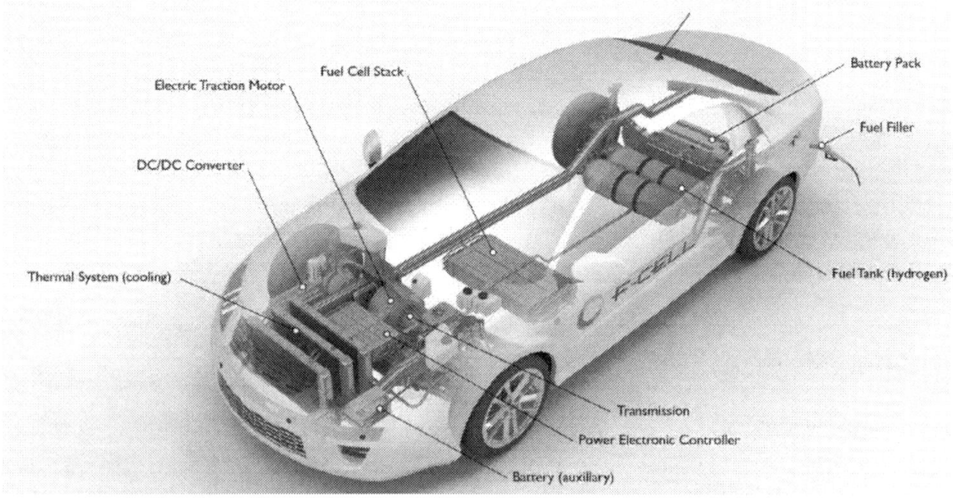

Figure 1: Fuel cell vehicle with on-board storage [29].

Hydrogen Uptake in Metal-Based Compounds

Metal hydration can be employed to store hydrogen below 3 or 4 MPa at above room temperature; however, the metals induce too much additional weight for most vehicles and are also expensive [33]. It has been found that lithium nitride can reversibly store large amounts of hydrogen. This material stores hydrogen rapidly in the temperature range of 170–210 °C and obtained a 9.3 wt% uptake when the sample was held at 255 °C for 30 min. Under high vacuum (10–9 MPa, 10–5 mbar) about two-thirds of the hydrogen was released at temperatures below 200 °C. The remaining third of the stored hydrogen required temperatures above 320 °C for release. The hydrogen was taken up as lithium amide (LiNH2) and lithium hydride (LiH). The authors suggested that the related metal-N-H systems should be studied to find more practical pressures and temperatures for a hydrogen storage system [34].

Cryogenic Liquid Hydrogen Storage

In the cryogenic liquid storage method, hydrogen is stored in the liquid form by getting to the cryogenic temperature of −259.2 °C. Liquid hydrogen (LH_2) is low density, and 1 L of liquid hydrogen weighs only 71.37×10^{-3} kg. One liter of hydrogen can produce electricity of 8.52 MJ. Maintaining the hydrogen at such a low temperature is highly challenging and requires insulation, thus increasing the cost. Liquid hydrogen becomes explosive when it mixes with certain other gases, thus before refilling the tank with hydrogen, one should use nitrogen gas to drain the residual gases present inside the tank [35–37].

PRINCIPLES OF FUEL CELL

There are various types of FC systems. However, the principle of their function is similar. For a fuel cell system, three pillars are required: an anode, a cathode, and an electrolyte. FCs are categorized by the type of electrolyte material used. An FC can be composed of hundreds of individual cells, but each has the three same fundamental components. The electrolyte is located between the cathode and the anode. Figure 2 depicts a schematic of a polymer electrolyte FC (PEMFC) operation diagram [38]. This FC type is also known as a proton exchange membrane FC. The PEMFC is what is most commonly used in mobile power applications, such as vehicles. While the electrolyte material used varies depending on the type of FC, the general function of the FC is as follows—fuel (pure hydrogen) is fed into the anode compartment of the fuel cell while air or pure oxygen is fed into the cathode side of the FC. On the anode side of the cell, electrons are separated as the gas tries to make its way through the electrolyte membrane. The membrane acts as a filter to separate the electrons and the hydrogen ions while only allowing the hydrogen ions to pass through. In the cathode compartment, the hydrogen ions that passed through the membrane combine with the oxygen atoms from the air supply to produce H_2O as a by-product; heat is also produced as a by-product [39]. Unlike internal combustion engines, where the fuel is mixed with air and fuel, there is separation of the fuel and the oxidant with no combustion of the fuel in an FC. Therefore, FCs do not produce the harmful emissions that internal combustion engines produce.

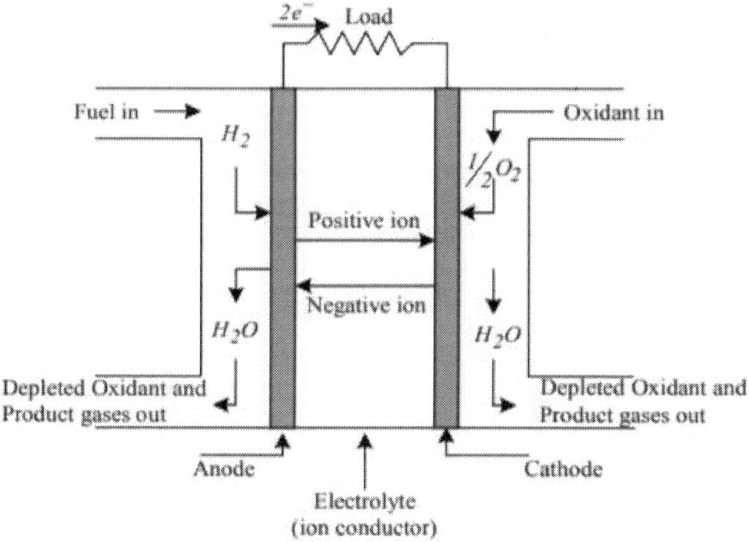

Figure 2: Fuel cell operation diagram [38].

Types of Fuel Cell

FC systems are classified by the type of membrane they use. Table 1 shows some of the more common FCs and the type of membrane each uses.

Table 1: Classification of fuel cell systems based on the employed membrane [40].

Fuel Cell	Abbreviation	Membrane
Solid Oxide Fuel cells	SOFC	Yttria-stabilized zirconia
Direct methanol fuel cell	DMFC	Solid polymer electrolyte (Nafion)
Phosphoric Acid fuel cell	PAFC	Phosphoric Acid (H_3PO_4)
Polymer electrolyte fuel cell Or Proton exchange membrane	PEMFC	Solid polymer electrolyte (Nafion)
Alkaline fuel cell	AFC	Aqueous solution Potassium Hydroxide (KOH)

Solid Oxide Fuel Cell (SOFC)

A common stationary power FC is a solid oxide fuel cell [41]. SOFCs conduct ions in a ceramic membrane at high temperatures. In SOFCs, the ceramic usually is a yttrium stabilized zirconia (YSZ) that will conduct oxygen ions (O_2), but other ceramics conduct hydrogen ions. Solid oxide fuel cells operate at elevated temperature ranges, usually around 650–800 °C. An advantage to an SOFC is that, due to the elevated temperature ranges and oxidizing ion (O^-) transport, these fuel cells can handle a variety of fuels [39]. A common use for these fuel cells is as backup power for cell phone towers. The reactions associated with solid oxide fuel cells are presented below [42].

Full reaction:

$$H_2 + \frac{1}{2}O_2 \rightarrow H_2O \quad ...1$$

Oxidation reactions (anode):

$$H_2 + O^{2-} \rightarrow H_2O + 2e^- \quad ...2$$

CO and hydrocarbons can also be used in SOFC at high temperatures, thus a water gas shift reaction is possible:

$$CO + H_2O \rightarrow H_2 + CO_2 \quad ...(3)$$

and the steam reforming reaction is also possible:

$$CH_4 + H_2O \rightarrow H_2 + CO_2 \text{ (in the case of natural gas separation)} \quad ...(4)$$

Reduction reaction (cathode):

$$H_2 + O^{2-} \rightarrow H_2O + 2e^- \quad ...(5)$$

Direct Methanol Fuel Cell (DMFC)

Methanol can be directly employed as fuel in the FC in a DMFC system. Methanol is an organic fuel produced from coal or agricultural products. In DMFCs, both the cathode and the anode are platinum or platinum-adopted catalysts. The electrolyte solution used is trifluromethane sulfonic acid. A DMFC is an example of a low temperature non-fuel flexible fuel cell. These fuel cells were initially implemented in small portable electronic devices, such as laptops and cell phones [43]. Compared to the PEMFC, the DMFC has lower density and efficiency.

Phosphoric Acid Fuel Cell (PAFC)

PAFCs rely on an acidic electrolyte, as with PEMFCs, to conduct hydrogen ions. The reactions in the anode and the cathode are the same as the PEMFC reactions. Phosphoric acid (H_3PO_4) is a viscous liquid that is contained by capillarity in the FC in a porous silicon carbide matrix. PAFCs are medium temperature fuel cells that conduct hydrogen ions, thus they are not as fuel-flexible as the high-temperature fuel cells that conduct oxidizing ions (e.g., O^-, CO_3^-). While PAFCs are predominantly used for stationary power, they have also been implemented in some large-scale vehicles, such as public buses [44]. The overall reaction can be characterized as follows:

$$H_2 + \frac{1}{2}O_2 \rightarrow H_2O \quad ...(6)$$

The half reaction associated with the phosphoric acid fuel cell is as follows:

Oxidation reaction:

$$H_2 \rightarrow 2H^+ + 2e^- \quad ...(7)$$

Reduction reaction:

$$\frac{1}{2}O_2 + 2H^+ + 2e^- \rightarrow H_2O \quad ...(8)$$

Polymer Electrolyte Membrane Fuel Cell (PEMFC)

Solid polymer membranes are used as the electrolyte in PEMFCs. The polymer membrane is perfluorosulfonic acid referred to as Nafion. This polymer membrane is acidic; hence, the ions transported are hydrogen ions or protons. The PEMFC is fueled with pure hydrogen, and the oxidant is air or pure oxygen. PEMFCs are low temperature fuel cells that conduct hydrogen ions (H^+), making them not fuel-flexible. These fuel cells are the most widely used in the transportation sector because they are low temperature FCs, operating around 80 °C, thus they have relatively short starting and stopping times. Another advantage to PEMFCs is that they have very high efficiency and power density in the vehicle engine size class [45]. These features are well-suited to a vehicle power source where power density is desired and the dynamic power demands are significant. Some of the drawbacks to PEMFCs include a higher risk of CO poisoning and that they require cooling.

The overall chemical reaction for PEMFC is:

$$H_2 \rightarrow 2H^+ + 2e^- \quad ...(9)$$

The chemical half reactions for the PEMFC are as follows:

Oxidation reaction:

$$H_2 \rightarrow 2H^+ + 2e^- \quad ...(10)$$

Reduction reaction:

$$\frac{1}{2}O_2 + 2H^+ + 2e^- \rightarrow H_2O \quad ...(11)$$

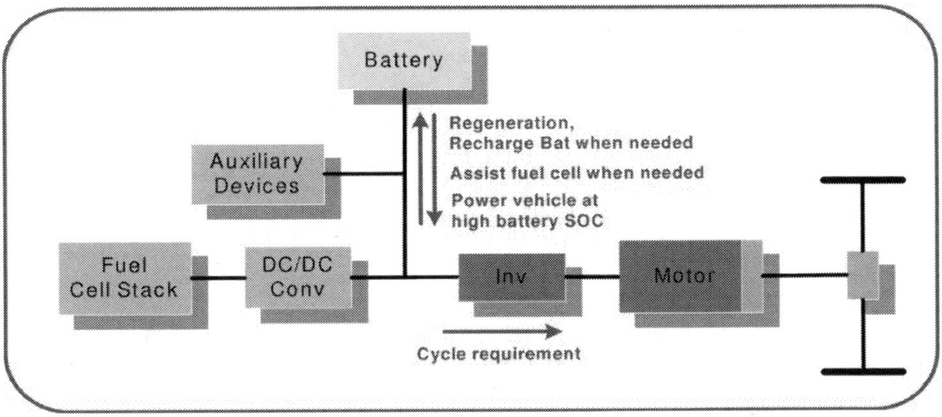

Figure 3: Basic structure of fuel cell based electric vehicle [47].

Different configurations of the hybrid electric vehicle are discussed in the following section.

Alkaline Fuel Cells (AFC)

AFCs use an aqueous solution of potassium hydroxide (KOH) as the electrolyte to conduct ions between electrodes. Since the electrolyte is alkaline, the ion conduction mechanism is different from PEMFCs. The ion carried by the alkaline electrolyte is a hydroxide ion (OH), which affects several other aspects of the FC.

The half reactions are as follows:

Oxidation:

$$H_2 + 2\,OH^- \rightarrow 2H_2O + 2e^- \quad \text{...(12)}$$

Reduction:

$$O2 + 4e^- + 2H_2O \rightarrow 4\,OH^- \quad \text{...(13)}$$

Water is required at the cathode by the oxygen reduction, thus water management is an important issue that is sometimes resolved by using waterproof electrodes and keeping the water in the electrolyte. The cathode reaction uses water from the electrolyte, whereas the anode reaction rejects its produced water. The excess water (2 mol per reaction) is evaporated outside the stack. AFCs can operate over a wide range of pressures and temperatures, from 0.22 to 4.5 MPa and 80 to 230 °C, respectively. High-temperature AFCs also make use of a highly concentrated electrolyte— highly concentrated enough that the ion transport mechanism changes from an aqueous solution to molten salt [48].

Unitized Reversible Fuel Cell

A unitized reversible FC (URFC) is an energy-storage device that performs in water-electrolysis mode (EC mode) to produce hydrogen and works in FC mode to generate electricity. During the mode switching of the URFC, a reversible electrochemical reaction takes place that causes the change in temperature [49]. Research on URFCs found pre-reactant switching, oxygen flow rate, hydrogen flow rate, and time interval length makes the URFC more efficient and reliable. The water accumulation in URFCs is a major problem, as it decreases the mass flow rate of reactants in FC mode. It also affects the mode switching in URFC. The residual water left during EC mode and the water produced during FC mode should be eliminated to facilitate smooth mode switching. Gas purging has been used to remove the water at the proton exchange membrane at the end of the FC operation [50]. Gas purge time increases with a decrease in the cell temperature. The gas purge flow rate should be greater than the purging time. Due to the high flow rates, the water droplets in the PEM can be pulled away from it. In FC mode, the mass flow rate of reactants is affected by the water content in the cell. To make the successful mode switching process from electrolysis mode to fuel cell mode, enough time should be provided. Allowing adequate time for gas purging enhances the mass flow rate, which helps the FC start up. The water in the channel is pushed by the oxygen gas, but the water in the oxygen side is still present due to the increased amount of time required for water electrolysis to occur. Pre-reactant switching is the method where the reactant gases (oxygen and hydrogen) are switched on before transitioning to the FC mode. By supplying the oxygen to the FC before the current supply, the residual water stored in the channels and the gas diffusion layer made during EC mode is eliminated. The gas must be supplied 180 s before the current transition to FC mode [51]. This will effectively consume the residual water at end of EC mode.

FUEL CELL HYBRID VEHICLES

The development of one important aspect of the fuel cell vehicle—the electric motor—dates back to the early 19th century. Although electric vehicles were a strong contender in the early 20th century to become a mainstream transportation method, the ICE vehicle eventually won out due to the short range and the high cost of electric vehicles. Additionally, the discovery of Texan oil reduced the price of gasoline, thus it became affordable to the average consumer, which caused fuel cell and electric vehicles to take a backseat to the ICE vehicle for most of the last 100 years [47]. The oil embargo in 1973 kick-started a renewed interest in FC power for personal transportation applications, as governments looked to mitigate their dependence on petroleum imports. In the early 1970s, K. Kordesch [52] modified a sedan to operate from a 6-kW FC and a lead acid battery pack. The automobile was driven on public roads for about three years. In 1993, Ballard launched a fuel cell-powered light-duty transit bus using a 120-kW FC system, followed by a heavy-duty transit bus using a 200-kW FC system in 1995. In 1994 and 1995, H-Power built three fuel cell-battery hybrid buses, each using a 50-kW FC and a 100-kW nickel-cadmium battery [53]. The importance of these releases was to make FC technology understandable to key decision makers in industry and government. These buses helped to prove that fuel cells would work in the real world. Fleet-vehicle operations, such as buses and delivery services, were early adopters of fuel cell technology due to the ease of centralized refueling and the reduced requirement for a high range capability between fill-ups. Trials of FC powered buses have occurred in Vancouver and Chicago, as well as in other cities in Europe and North America [54]. In July 2005, the first FC vehicle was leased to a family in California as an important step in getting more fuel cell vehicles on the road. However, many obstacles remain to be overcome before the FC vehicle can become a mainstream form of transportation. An obvious issue is the need for a hydrogen infrastructure to enable refueling of the vehicles. Hydrogen filling stations currently exist in many countries around the world, such as Canada, the USA, Iceland, Japan, Singapore, and Germany. Though these stations are currently not widespread enough to allow large numbers of people to begin driving FCVs, it is expected that more hydrogen infrastructure will be built as more FCVs become commercially available. Hydrogen FCVs have evolved significantly; currently, they can drive between 311 to 597 miles on a full tank. The development of these vehicles is increasing, but they still require significant improvements [14].

It should be noted that FCEVs are more promising in city bus applications due to two reasons—the supply of hydrogen is not crucial, because the buses refuel in one place, hence only one refilling point is required, and the price of FCs. FCs are still expensive, therefore it makes more sense to buy them for vehicles that are in use for many hours each day [55]. As of June 2018, there have been more than 6500 FCVs delivered to consumers. California was the leading market for FCVs, with nearly 3000 vehicles being delivered due to it having the largest network of hydrogen refueling stations (HRS) [56]. In Europe, Germany's Linde AG and France's Air Liquide have been working together to increase Germany's stations from 15 to 100 by 2017 and to 400 by 2023. About 1000 HRS would be required to provide full coverage in countries such as Germany or France, with a cost of 1.5 to 2 billion euro [57,58]. FCV sales volumes are projected to be significant, but only in the long term, even with a favorable climate-policy scenario. Figure 4 shows FCV sales volume anticipations based on a long-term powertrain mix scenario (million annual units) [59]. Considering a similar scenario, the international energy agency (IEA) anticipates an FCV market share of about 17% by 2050 (35 million annual unit sales) [60].

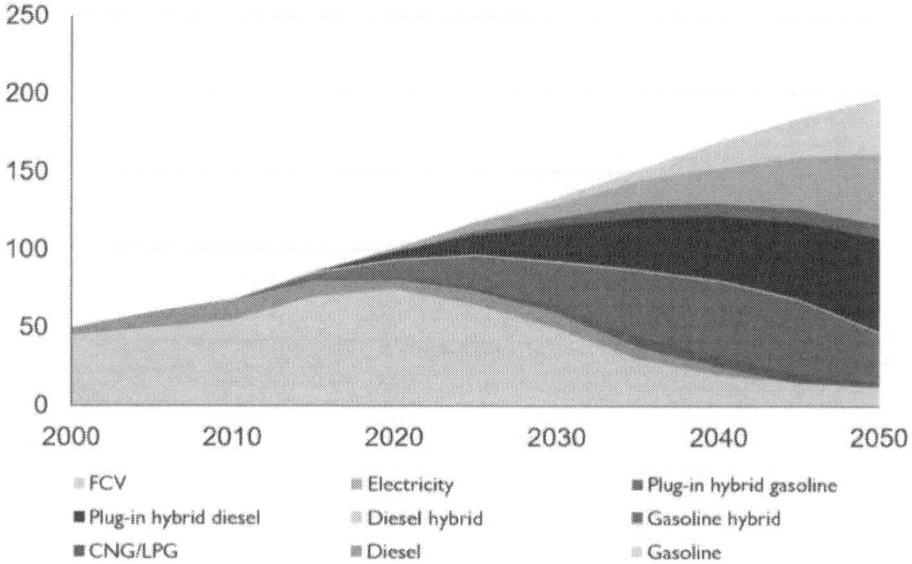

Figure 4: Fuel cell vehicle (FCV) sales volume. Source: IEA. All rights reserved. [59].

Another option to increase the HRS prevalence is to use electrolysis at refueling stations to convert electricity from the grid into hydrogen. This idea could be extended to residential applications, where people would have a hydrogen refueling station in their own homes [61–63]. The obstacles associated with developing an adequate hydrogen infrastructure brings up another important question—where will the hydrogen come from? Although hydrogen is the most abundant element in the universe, it rarely exists alone in nature. Today, hydrogen is mostly produced by reforming natural gas. In this way, pollutants can be captured if the reforming is done at a central plant. However, other options, such as direct solar hydrogen from methane at landfills and hydrogen from bacteria [64], are continually being explored. If electrolysis is used to generate hydrogen, there may have to be an increase in electricity generation to satisfy the need for hydrogen, though the increase may be small, since hydrogen can be produced during off-peak times (e.g., overnight). Methods to generate electricity that have a minimal impact on the environment include nuclear, wind, solar, hydro, and geothermal. It is often said that, at the beginning of the hydrogen economy, most of the hydrogen will still be reformed from natural gas, but as time goes by, society will move towards more ideal sources of energy, such as wind and solar [65]. Although generation and transportation of hydrogen is a major issue in the deployment of fuel cell vehicles, many diverse ideas are being developed to solve the problem. Other obstacles for fuel cell vehicles include improving on-board hydrogen storage, improving fuel cell and battery durability, and increasing the efficiency and the performance of a fuel cell vehicle. The last barrier to commercialization is cost—a true challenge—as fuel cells are still extremely expensive today. Many parts used to work with fuel cells (e.g., power electronic converters) must be custom-made and can be very expensive as a result [66]. However, these costs will decrease as technological processes are improved and components are mass-produced.

Parallel Hybrid

One of the earlier hybrid drive train designs was the parallel hybrid architecture. In this configuration, both the internal combustion engine and the electric motor can be used to power the vehicle independent of one another. The ICE has a low torque output at low speeds, therefore in stop-and-go drive cycles, the ICE is extremely inefficient. However, electric motors provide almost instantaneous torque, making them ideal for stop-and-go drive cycles. Implementing a parallel hybrid system allows each power source or both to operate at varying degrees when it is most efficient in the drive cycle [35]. For example, during low speeds and stop-and-go cycles, the electric motor would primarily be used, and the internal combustion engine would be mostly off for efficiency reasons. Conversely, during steady-state highway conditions, the electric motor would be off, while the ICE would power the vehicle to induce a long driving range by utilizing the high energy density of the fuel [65]. In the parallel configuration, one can also operate both the ICE and the electric motor together when the torque demand is greater than the torque produced by either the ICE or the electric motor independently [66]. Since both power sources can be operated in unison when torque demands are high, it allows both the engine and the motor to be scaled down in size, since most of the time the drive cycle will not have a high torque demand. Reducing the size of both motors can further increase the overall efficiency of the parallel design. Another advantage to the parallel hybrid drive train architecture is the ability to harvest energy through deceleration in the drive cycle with the implementation of regenerative braking. Regenerative braking is not exclusive to parallel hybrid architecture, rather, almost all hybrids can deploy regenerative braking to harvest what would otherwise be wasted heat energy from the mechanical brakes [67]. Regenerative braking works by converting the electric motor into a generator during stops and decelerations in the drive cycle; this energy then gets stored in the batteries of the vehicle to be used later in the drive cycle [68]. A visual representation of a parallel hybrid drive train is depicted below in Figure 5 [69].

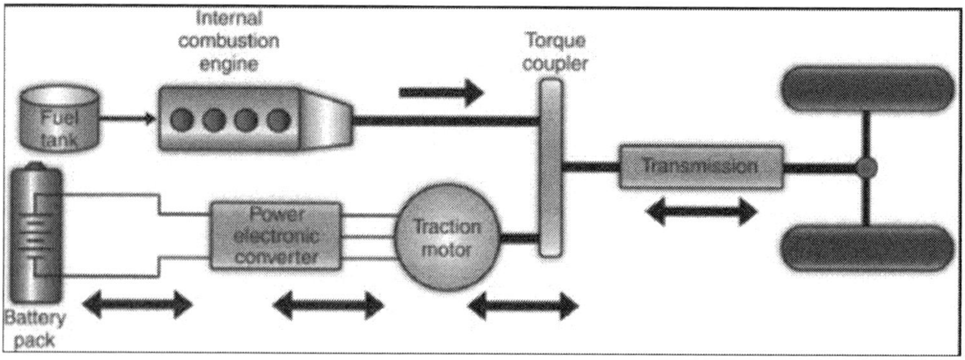

Figure 5: Parallel hybrid configuration [69].

Series Hybrid

A series hybrid drive train architecture differs from a parallel hybrid drive train in that the two power sources are no longer independently able to power the vehicle. In this configuration, generally the ICE will act as a charger for the batteries that supply electricity to the electric motor—the only source to propel the vehicle [70]. One of the advantages of this design is that there is no need for a transmission, since the ICE is not powering the vehicle directly. Eliminating the transmission from the engine system reduces the weight of the vehicle, which directly correlates to increased efficiency. In a series configuration, the ICE does not have to account for any of the transient dynamics in the drive cycle as it would in a parallel configuration. Because of this, the engine can be operated at a steady state at its most efficient rpm, further increasing its efficiency. In a series configuration, there is more flexibility regarding the implementation of alternative energy devices. For example, for this project, the ICE was replaced with a hydrogen FC. In a series hybrid, the ICE converts chemical energy to mechanical energy and then to electrical energy, thus it is actually more efficient to use an FC to convert directly from chemical energy to electrical energy to charge the battery pack. This is one of the main reasons the decision was made to implement a series configuration in this project. It should be noted that a fuel cell can be used in a parallel configuration as well. However, to achieve parallel use, the fuel cell would need to be robust enough to handle the transient dynamics created by the drive cycle. Most fuel cells have high enough power density to handle transient dynamic loading well. Usually, fuel cells are required to operate near steady-state conditions, which makes them ideal for series hybrid architectures [71–75]. An example of a series hybrid configuration with an ICE is depicted below in Figure 5.

The series hybrid design can also support regenerative braking to recover energy during stops and decelerations the same way the parallel hybrid does by converting the motor into a generator during these instances in the drive cycle. A drawback to the series hybrid design is that the motor is required to be larger, since there is no way to use both power sources during a hard acceleration.

Series-Parallel Hybrid

The series-parallel hybrid is a combination of both series and parallel hybrid drive train configurations. In a series-parallel hybrid, both the electric motor and the ICE can power the vehicle independently or jointly. The ICE can divide its power between driving the car and charging the batteries simultaneously [76]. This configuration provides the most control flexibility, since the modes of operation can be optimized to compensate for the current drive cycle. However, one of the drawbacks to this design is the amount of hardware required. Since the ICE must be able to drive the vehicle, a transmission is required, and a generator is required to enable the ICE to charge the batteries. The series-parallel design has all the benefits of both the series design and the parallel design, which makes it more flexible to different drive cycles; however, it also has all the drawbacks of both designs [77–81]. An example of a series-parallel hybrid configuration is shown below in Figure 6.

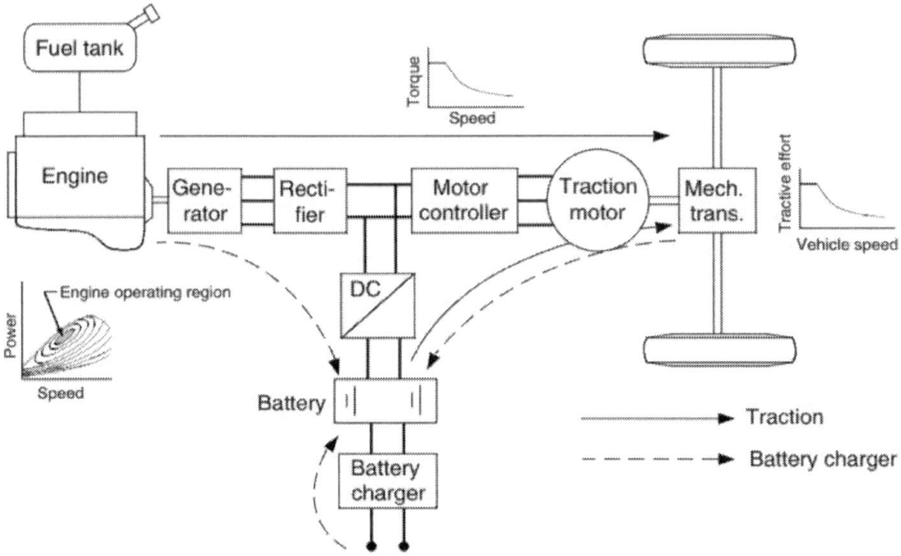

Figure 6: Series hybrid configuration [35].

CONTROL STRATEGIES TO MAKE FCV MORE EFFICIENT

The disadvantage of FCV, such as low power density and low power response, can be overcome by implementing super capacitors (SCAPS) and batteries (BATs) with a fuel cell in an FCV [82–84]. An FCEV has a three phase traction motor, an energy storage system, a direct current (DC) bus, and auxiliary devices. The energy is transferred in the order of FC-BAT-SCAP, and an inverter is used to control the output energy to the traction motor and the output from the SCAP. The most common control strategies used by these controllers are given below.

1. Peaking Power Source Strategy (PPSS)
2. Operating Mode Control Strategy (OMCS)
3. Fuzzy Logic Control Strategy (FLCS)
4. Equivalent Consumption Minimization Strategy (ECMS)

Peaking Power Source Strategy (PPSS)

In PPSS, the controllers, the wheels, the traction motor, and the pedals are electronically interfaced to collect the signal. The power and the torque are controlled by the vehicle controller based on the commands from the gas and the brake pedals. Based on the demanded power, the energy alters between FC and PPS to meet the vehicle power requirement. The required energy is provided either from the FC, with the PPS, or both, and according to the demand from the driver to respond quickly, traction is designed to control the output power [85,86].

Operating Mode Control Strategy (OMCS)

OMCS manages the power variances and the power control between the FC and the BAT. In the discharge mode, the vehicle's power demand (P_{demand}) may be less or more than the nominal P_{FC}, depending on the acceleration. If P_{demand} is lower than the nominal P_{FC}, the required energy is provided by FC. If P_{demand} is more than the nominal P_{FC}, the required energy is generated by both FC and from BAT. In charging mode, P_{demand} is the nominal PFC, and the FC both provides the power demand and charges the BAT. P_{demand} is much smaller than the nominal P_{FC} in the fast charge mode. In this condition, the FC supplies both P_{demand} and the power to all units that require extra energy [87,88].

Fuzzy Logic Control Strategy (FLCS)

To control the energy storage system and enhance the efficiency of the FCV, FLCS is applied. While the power levels of SCAP and BAT state of charging (SOC) are low (L_{SOC}), P_{DC} ought to be maximized (DC_{max}). When SCAP and BAT power levels are high, P_{DC} must either be at the minimum level (DC_{min}) or the middle level (DC_{middle}), depending on P_{demand}. The goal is to maintain the optimum power levels of SCAP and BAT while meeting P_{demand}, allowing operation at a nominal P_{FC}.

Equivalent Consumption Minimization Strategy (ECMS)

The ECMS is applied to manage the energy resource in the energy storage device. The electricity consumption of SCAP and BAT is converted into the equivalent hydrogen consumption. Moreover, ECMS can use electrical energy supplied by energy storage systems such as the SCAP and the BAT as hydrogen from the FC, if necessary, with the concept of equivalent fuel consumption [89].

CONCLUSIONS

In this paper, an investigation into hydrogen-based energy generation using fuel cells and their utilization in hybrid vehicles was conducted. Various types of FCs and their applications were analyzed, namely the application of FCs in hybrid vehicles. FCs with adequate control strategies could become the more advantageous choice for applications in vehicles over batteries alone. Hydrogen fuel cells will play a significant role in the transportation industry in the near future. The price of fuel cells will reduce when producing fuel cells in large quantities and commercializing them. We could expect fuel cell-based transportation, power plants, and electricity generators to become prominent in the coming decades.

Author Contributions:

Conceptualization, S.E.H., writing—original draft preparation, Y.M., B.B., H.A., B.T.F.S.; writing—review and editing, T.A., J.K.

Funding:

This research received no external funding.

Conflicts of Interest:

The authors declare no conflict of interest.

ABBREVIATION

AFC	Alkaline Fuel Cell
BAT	Battery
BEV	Battery electric vehicle
CH2	Compressed Hydrogen
CV	Conventional vehicle
DMFC	Direct methanol fuel cell
ECMS	Equivalent Consumption Minimization Strategy
FC	Fuel cell
FCEV	Fuel cell electric vehicle
FCV	Fuel cell vehicle
FLCS	Fuzzy Logic Control Strategy
HRS	Hydrogen refueling stations
ICE	Internal combustion engine
LH2	Liquid Hydrogen
OMCS	Operating Mode Control Strategy
PAFC	Phosphoric acid fuel cell
PEMFC	Polymer electrolyte membrane fuel cell
PPSS	Peaking Power Source Strategy
SCAP	Super Capacitor
SOC	State of Charge
SOFC	Solid oxide fuel cells
URFC	Unitized regenerative fuel cell
YSZ	Yttrium stabilized zirconia

REFERENCES

Hosseini, S.E.; Andwari, A.M.; Wahid, M.A.; Bagheri, G. A review on green energy potentials in Iran. *Renew. Sustain. Energy Rev.* **2013**, *27*, 533–545. [CrossRef]

Granovskii, M.; Dincer, I.; Rosen, M.A. Greenhouse gas emissions reduction by use of wind and solar energies for hydrogen and electricity production: Economic factors. *Int. J. Hydrogen Energy* **2007**, *32*, 927–931. [CrossRef]

Derbeli, M.; Barambones, O.; Sbita, L.; Derbeli, M.; Barambones, O.; Sbita, L. A Robust Maximum Power Point Tracking Control Method for a PEM Fuel Cell Power System. *Appl. Sci.* **2018**, *8*, 2449. [CrossRef]

Hosseini, S.E.; Wahid, M.A.; Aghili, N. The scenario of greenhouse gases reduction in Malaysia. *Renew. Sustain. Energy Rev.* **2013**, *28*, 400–409. [CrossRef]

Eriksson, E.L.V.; Gray, E.M.A. Optimization and integration of hybrid renewable energy hydrogen fuel cell energy systems—A critical review. *Appl. Energy* **2017**, *202*, 348–364. [CrossRef]

Heywood, J.B. *Internal Combustion Engine Fundamentals*; McGraw-Hill Education: New York, NY, USA, 1988.

Costilla-Reyes, A.; Erbay, C.; Carreon-Bautista, S.; Han, A.; Sánchez-Sinencio, E. A Time-Interleave-Based Power Management System with Maximum Power Extraction and Health Protection Algorithm for Multiple Microbial Fuel Cells for Internet of Things Smart Nodes. *Appl. Sci.* **2018**, *8*, 2404. [CrossRef]

Kerviel, A.; Pesyridis, A.; Mohammed, A.; Chalet, D.; Kerviel, A.; Pesyridis, A.; Mohammed, A.; Chalet, D. An Evaluation of Turbocharging and Supercharging Options for High-Efficiency Fuel Cell Electric Vehicles. *Appl. Sci.* **2018**, *8*, 2474. [CrossRef]

Wang, C.; Nehrir, M.H.; Gao, H. Control of PEM Fuel Cell Distributed Generation Systems. *IEEE Trans. Energy Convers.* **2006**, *21*, 586–595. [CrossRef]

Somekawa, T.; Nakamura, K.; Kushi, T.; Kume, T.; Fujita, K.; Yakabe, H. Examination of a high-efficiency solid oxide fuel cell system that reuses exhaust gas. *Appl. Therm. Eng.* **2017**, *114*, 1387–1392. [CrossRef]

Ayad, M.Y.; Becherif, M.; Henni, A. Vehicle hybridization with fuel cell, supercapacitors and batteries by sliding mode control. *Renew. Energy* **2011**, *36*, 2627–2634. [CrossRef]

Chakraborty, U.K. A New Model for Constant Fuel Utilization and Constant Fuel Flow in Fuel Cells. *Appl. Sci.* **2019**, *9*, 1066. [CrossRef]

Giorgi, L.; Leccese, F. Fuel Cells: Technologies and Applications. *Open Fuel Cells J.* **2013**, *6*, 1–20. [CrossRef]

Offer, G.J.; Howey, D.; Contestabile, M.; Clague, R.; Brandon, N.P. Comparative analysis of battery electric, hydrogen fuel cell and hybrid vehicles in a future sustainable road transport system. *Energy Policy* **2010**, *38*, 24–29. [CrossRef]

Choi, H.; Shin, J.; Woo, J. Effect of electricity generation mix on battery electric vehicle adoption and its environmental impact. *Energy Policy* **2018**, *121*, 13–24. [CrossRef]

Aceves, S.M.; Berry, G.D.; Weisberg, A.H.; Espinosa-Loza, F.; Perfect, S.A. Advanced concepts for vehicular containment of compressed and cryogenic hydrogen. In Proceedings of the 16th World Hydrogen Energy Conference 2006 (WHEC 2006), Lyon, France, 13–16 June 2006.

Michel, F.; Fieseler, H.; Allidiers, L. Liquid Hydrogen Technologies for Mobile Use. In Proceedings of the 16th World Hydrogen Energy Conference 2006 (WHEC 2006), Lyon, France, 13–16 June 2006.

Paggiaro, R.; Michl, F.; Bénard, P.; Polifke, W. Cryo-adsorptive hydrogen storage on activated carbon. II: Investigation of the thermal effects during filling at cryogenic temperatures. *Int. J. Hydrogen Energy* **2010**, *35*, 648–659. [CrossRef]

Hosseini, S.E.; Wahid, M.A. Hydrogen production from renewable and sustainable energy resources: Promising green energy carrier for clean development. *Renew. Sustain. Energy Rev.* **2016**, *57*. [CrossRef]

Boretti, A. Novel dual fuel diesel-ammonia combustion system in advanced TDI engines. *Int. J. Hydrogen Energy* **2017**, *42*, 7071–7076. [CrossRef]

Veziroglu, A.; MacArio, R. Fuel cell vehicles: State of the art with economic and environmental concerns. *Int. J. Hydrogen Energy* **2011**, *36*, 25–43. [CrossRef]

Hosseini, S.E.; Wahid, M.A.; Ganjehkaviri, A. An overview of renewable hydrogen production from thermochemical process of oil palm solid waste in Malaysia. *Energy Convers. Manag.* **2015**, *94*, 415–429. [CrossRef]

Hosseini, S.E.; Abdul Wahid, M.; Jamil, M.; Azli, A.A.M.; Mohamad, F.M. A review on biomass-based hydrogen production for renewable energy supply. *Int. J. Energy Res.* **2015**, *39*, 1597–1615. [CrossRef]

Hibino, T.; Kobayashi, K.; Ito, M.; Ma, Q.; Nagao, M.; Fukui, M.; Teranishi, S. Efficient Hydrogen Production by Direct Electrolysis of Waste Biomass at Intermediate Temperatures. *ACS Sustain. Chem. Eng.* **2018**, *6*, 9360–9368. [CrossRef]

Hibino, T.; Kobayashi, K.; Lv, P.; Nagao, M.; Teranishi, S.; Mori, T. An Intermediate-Temperature Biomass Fuel Cell Using Wood Sawdust and Pulp Directly as Fuel. *J. Electrochem. Soc.* **2017**, *164*, F557–F563. [CrossRef]

Hibino, T.; Kobayashi, K.; Ito, M.; Nagao, M.; Fukui, M. Applied Catalysis B: Environmental Direct electrolysis of waste newspaper for sustainable hydrogen production: An oxygen-functionalized porous carbon anode. *Appl. Catal. B Environ.* **2018**, *231*, 191–199. [CrossRef]

Hibino, T.; Kobayashi, K.; Nagao, M.; Teranishi, S. Hydrogen Production by Direct Lignin Electrolysis at Intermediate Temperatures. *ChemElectroChem* **2017**, *4*, 3032–3036. [CrossRef]

Hori, T.; Ma, Q.; Kobayashi, K.; Nagao, M. Electrolysis of humidified methane to hydrogen and carbon dioxide at low temperatures and voltages. *Int. J. Hydrogen Energy* **2019**, *44*, 2454–2460. [CrossRef]

Alternative Fuels Data Center: How Do Fuel Cell Electric Vehicles Work Using Hydrogen? Available online: https://afdc.energy.gov/vehicles/how-do-fuel-cell-electric-cars-work (accessed on 21 May 2019).

Devrim, Y.; Devrim, H.; Eroglu, I. Development of 500 W PEM fuel cell stack for portable power generators. *Int. J. Hydrogen Energy* **2015**, *40*, 7707–7719. [CrossRef]

Davis, C.; Edelstein, B.; Evenson, B.; Brecher, A.; Cox, D. Hydrogen Fuel Cell Vehicle Study. Presented at the Panel on Public Affairs (POPA), American Physical Society, 12 June 2003.

Atkinson, K.; Roth, S.; Hirscher, M.; Grünwald, W. Carbon nanostructures: An efficient hydrogen storage medium for fuel cells. *Fuel Cells Bull.* **2001**, *4*, 9–12. [CrossRef]

Das, L.M. Hydrogen-fueled internal combustion engines. *Compend. Hydrog. Energy* **2016**, 177–217. [CrossRef]

Chen, P.; Xiong, Z.; Luo, J.; Lin, J.; Tan, K.L. Interaction of hydrogen with metal nitrides and imides. *Nature* **2002**, *420*, 302–304. [CrossRef]

Ehsani, M.; Gao, Y.; Longo, S.; Ebrahimi, K. *Modern Electric, Hybrid Electric, and Fuel Cell Vehicles, Third Edition*; CRC Press: Boca Raton, FL, USA, 2018; ISBN 9780429504884.

Jorgensen, S.W. Hydrogen storage tanks for vehicles: Recent progress and current status. *Curr. Opin. Solid State Mater. Sci.* **2011**, *15*, 39–43. [CrossRef]

Greene, D.L.; Leiby, P.N.; James, B.D.; Perez, J.; Melandez, M.; Milbrandt, A.; Unnasch, S.; Rutherford, D. *Analysis of the Transition to Hydrogen Fuel Cell Vehicles and the Potential Hydrogen Energy Infrastructure Requirements*; Oak Ridge National Lab. (ORNL): Oak Ridge, TN, USA, 2008.

Kirubakaran, A.; Jain, S.; and Nema, R.K. The PEM Fuel Cell System with DC / DC Boost Converter: Design, Modeling and Simulation. *Int. J. Recent Trends Eng.* **2009**, *1*, 157.

O'Hayre, R.P.; Cha, S.-W.; Colella, W.G.; Prinz, F.B. *Fuel Cell Fundamentals*; John Wiley & Sons: Hoboken, NJ, USA, 2016; ISBN 9781119113805.

Reschiotto, D. Effects of Anode Fuel Recirculation on SOFCs Fuelled with Biogas. Master's Thesis, Politecnico di Torino, Torino, Italy, 2018.

Yang, G.; Jung, W.; Ahn, S.-J.; Lee, D.; Yang, G.; Jung, W.; Ahn, S.-J.; Lee, D. Controlling the Oxygen Electrocatalysis on Perovskite and Layered Oxide Thin Films for Solid Oxide Fuel Cell Cathodes. *Appl. Sci.* **2019**, *9*, 1030. [CrossRef]

NFCRC Tutorial: Solid Oxide Fuel Cell (SOFC). Available online: http://www.nfcrc.uci.edu/3/TUTORIALS/EnergyTutorial/sofc.html (accessed on 16 October 2018).

Barbir, F. *PEM Fuel Cells: Theory and Practice*, 2nd ed.; Academic Press: Cambridge, MA, USA, 2013; ISBN 9780123877109.

Yamarone, R. *The Trader's Guide to Key Economic Indicators*; Bloomberg Press: Hoboken, NJ, USA, 2004; ISBN 1576601390.

Srinivasan, S. Fuel cells: From fundamentals to applications. In *Fuel Cells*; Springer Science & Business Media: Boston, MA, USA, 2006; pp. 3–25.

Wilberforce, T.; El-Hassan, Z.; Khatib, F.N.; Al Makky, A.; Baroutaji, A.; Carton, J.G.; Olabi, A.G. Developments of electric cars and fuel cell hydrogen electric cars. *Int. J. Hydrogen Energy* **2017**, *42*, 25695–25734. [CrossRef]

Briguglio, N.; Andaloro, L.; Ferraro, M.; Antonucci, V. Fuel Cell Hybrid Electric Vehicles. In *Electric Vehicles, The Benefits and Barriers*; InTech: London, UK, 2011.

Garche, J.; Jörissen, L. Applications of Fuel Cell Technology: Status and Perspectives. *Electrochem. Soc. Interface* **2015**, *24*, 39–43. [CrossRef]

Bhosale, A.C.; Mane, S.R.; Singdeo, D.; Ghosh, P.C. Modeling and experimental validation of a unitized regenerative fuel cell in electrolysis mode of operation. *Energy* **2017**, *121*, 256–263. [CrossRef]

Zhang, Y.; Zhang, H.; Ma, Y.; Cheng, J.; Zhong, H.; Song, S.; Ma, H. A novel bifunctional electrocatalyst for unitized regenerative fuel cell. *J. Power Sources* **2010**, *195*, 142–145. [CrossRef]

Ito, H.; Miyazaki, N.; Ishida, M.; Nakano, A. ScienceDirect Efficiency of unitized reversible fuel cell systems. *Int. J. Hydrogen Energy* **2016**, *41*, 5803–5815. [CrossRef]

Shukla, A.K.; Jackson, C.L.; Scott, K. The Promise of Fuel Cell-Based Automobiles. *Bull. Mater. Sci.* **2003**, *26*, 207–214. [CrossRef]

Stefan, J.; Minott, S.J.; Norman, R.; Scott, N.R. Feasibility of Fuel Cells for Energy Conversion on Dairy Farms. In Proceedings of the ASABE Annual Meeting, Sacramento, CA, USA, July 29–August 1 2001.

Chu, S.; Majumdar, A. Opportunities and challenges for a sustainable energy future. *Nature* **2012**, *488*, 294–303. [CrossRef]

Mahmoud, M.; Garnett, R.; Ferguson, M.; Kanaroglou, P. Electric buses: A review of alternative powertrains. *Renew. Sustain. Energy Rev.* **2016**, *62*, 673–684. [CrossRef]

Hardman, S.; Tal, G. Who are the early adopters of fuel cell vehicles? *Int. J. Hydrogen Energy* **2018**, *43*, 17857–17866. [CrossRef]

Schwelm, G. Hydrogenics fueling installations in Germany, France. *Fuel Cells Bull.* **2009**, *2009*, 7. [CrossRef]

Robledo, C.B.; Oldenbroek, V.; Abbruzzese, F.; van Wijk, A.J.M. Integrating a hydrogen fuel cell electric vehicle with vehicle-to-grid technology, photovoltaic power and a residential building. *Appl. Energy* **2018**, *215*, 615–629. [CrossRef]

International Energy Agency. Available online: http://www.iea.org/ (accessed on 20 January 2017).

Arena, F.; Spera, D.; Laguardia, F. What's in the Future for Fuel Cell Vehicles? Arthur D Little: Boston, MA, USA.

Hisschemöller, M.; Bode, R.; van de Kerkhof, M. What governs the transition to a sustainable hydrogen economy? Articulating the relationship between technologies and political institutions. *Energy Policy* **2006**, *34*, 1227–

1235. [CrossRef]

Brey, J.J.; Carazo, A.F.; Brey, R. Exploring the marketability of fuel cell electric vehicles in terms of infrastructure and hydrogen costs in Spain. *Renew. Sustain. Energy Rev.* **2018**, *82*, 2893–2899. [CrossRef]

Thomas, C.E. Fuel cell and battery electric vehicles compared. *Int. J. Hydrogen Energy* **2009**, *34*, 6005–6020. [CrossRef]

Barbosa, M.J.; Rocha, J.M.; Tramper, J.; Wijffels, R.H. Acetate as a carbon source for hydrogen production by photosynthetic bacteria. *J. Biotechnol.* **2001**, *85*, 25–33. [CrossRef]

Emadi, A.; Rajashekara, K.; Williamson, S.S.; Lukic, S.M. Topological Overview of Hybrid Electric and Fuel Cell Vehicular Power System Architectures and Configurations. *IEEE Trans. Veh. Technol.* **2005**, *54*, 763–770. [CrossRef]

Liu, Y.-L.; Tong, C.-C.; Jwo, W.-S.; Lin, S.-J. Design an Intelligent Neural-Fuzzy Controller for Hybrid Motorcycle. In *NAFIPS 2007. 2007 Annual Meeting of the North American Fuzzy Information Processing Society*; IEEE: San Diego, CA, USA, 24–27 June 2007; pp. 283–288.

Ahn, J.K.; Jung, K.H.; Kim, D.H.; Jin, H.B.; Kim, H.S.; Hwang, S.H. Analysis of a regenerative braking system for Hybrid Electric Vehicles using an Electro-Mechanical Brake. *Int. J. Automot. Technol.* **2009**, *10*, 229–234. [CrossRef]

Gao, Y.; Chen, L.; Ehsani, M. Investigation of the Effectiveness of Regenerative Braking for EV and HEV. In *SAE Techical Paper*; SAE International: Warrendale, PA, USA, 1999.

Arora, A.; Medora, N.K.; Livernois, T.; Swart, J. Safety of Lithium-Ion Batteries for Hybrid Electric Vehicles. *Electr. Hybrid Veh.* **2010**, 463–491. [CrossRef]

Walters, J.; Husted, H.; Rajashekara, K. Comparative Study of Hybrid Powertrain Strategies. *SAE Trans.* **2001**, *110*, 1944–1953.

Hogarth, W.H.J.; Benziger, J.B. Dynamics of Autohumidifed PEM Fuel Cell Operation. *J. Electrochem. Soc.* **2006**, *153*, A2139. [CrossRef]

Hasan, A.; Guo, S.; Wahab, M. Simulation of a proton exchange membrane fuel cell. *World J. Eng.* **2011**, *8*, 109–115. [CrossRef]

Guarnieri, M.; Moro, F. A novel circuit model of a proton exchange membrane fuel cell. *COMPEL Int. J. Comput. Math. Electr. Electron. Eng.* **2010**, *29*, 1562–1572. [CrossRef]

Kato, M.; Henry, A.; Graham, S.; Doan, D.H.; Fushinobu, K. Molecular dynamics simulation of oxygen transport characteristics in the electrolyte membrane of PEMFC. *Int. J. Numer. Methods Heat Fluid Flow* **2018**, *28*, 289–296. [CrossRef]

Electrochemical Society. *Journal of the Electrochemical Society*; Electrochemical Society: Pennington, NJ, USA, 1948.

Grammatico, S.; Balluchi, A.; Italia, D. A series-parallel hybrid electric powertrain for industrial vehicles. In *Proceedings of the Automotive Control View Project*; IEEE: Piscataway, NJ, USA, 2010.

Amodeo, S.J.; Chiacchiarini, H.G.; Solsona, J.A.; Busada, C.A. High-performance sensorless nonlinear power control of a flywheel energy storage system. *Energy Convers. Manag.* **2009**, *50*, 1722–1729. [CrossRef]

Xiong, W.; Zhang, Y.; Yin, C. Optimal energy management for a series–parallel hybrid electric bus. *Energy Convers. Manag.* **2009**, *50*, 1730–1738. [CrossRef]

Hu, X.; Murgovski, N.; Johannesson, L.; Egardt, B. Energy efficiency analysis of a series plug-in hybrid electric bus with different energy management strategies and battery sizes. *Appl. Energy* **2013**, *111*, 1001–1009. [CrossRef]

He, X.; Jiang, Y. Review of hybrid electric systems for construction machinery. *Autom. Constr.* **2018**, *92*, 286–296. [CrossRef]

Dreier, D.; Silveira, S.; Khatiwada, D.; Fonseca, K.V.O.; Nieweglowski, R.; Schepanski, R. Well-to-Wheel analysis of fossil energy use and greenhouse gas emissions for conventional, hybrid-electric and plug-in hybrid-electric city buses in the BRT system in Curitiba, Brazil. *Transp. Res. Part D Transp. Environ.* **2018**, *58*, 122–138. [CrossRef]

Camara, M.B.; Gualous, H.; Gustin, F.; Berthon, A. Design and New Control of DC/DC Converters to Share Energy Between Supercapacitors and Batteries in Hybrid Vehicles. *IEEE Trans. Veh. Technol.* **2008**, *57*, 2721–2735. [CrossRef]

Bauman, J.; Kazerani, M. An Analytical Optimization Method for Improved Fuel Cell–Battery–Ultracapacitor Powertrain. *IEEE Trans. Veh. Technol.* **2009**, *58*, 3186–3197. [CrossRef]

Bauman, J.; Kazerani, M. A Comparative Study of Fuel-Cell–Battery, Fuel-Cell–Ultracapacitor, and Fuel-Cell–Battery–Ultracapacitor Vehicles. *IEEE Trans. Veh. Technol.* **2008**, *57*, 760–769. [CrossRef]

Thounthong, P.; Raël, S.; Davat, B. Energy management of fuel cell/battery/supercapacitor hybrid power source for vehicle applications. *J. Power Sources* **2009**, *193*, 376–385. [CrossRef]

Cipollone, R.; Di Battista, D.; Marchionni, M.; Villante, C. Model based Design and Optimization of a Fuel Cell Electric Vehicle. *Energy Procedia* **2014**, *45*, 71–80. [CrossRef]

Ning, Q.; Xuan, D.; Kim, Y. Modeling and control strategy development for fuel cell hybrid vehicles. *Int. J. Automot. Technol.* **2010**, *11*, 229–238. [CrossRef]

Bizon, N.; Thounthong, P. Fuel economy using the global optimization of the Fuel Cell Hybrid Power Systems. *Energy Convers. Manag.* **2018**, *173*, 665–678. [CrossRef]

Gao, L.; Dougal, R.A.; Liu, S. Power Enhancement of an Actively Controlled Battery/Ultracapacitor Hybrid. *IEEE Trans. Power Electron.* **2005**, *20*, 236–243. [CrossRef]

2

An Overview of Hydrogen as a Vehicle Fuel

H. Fayaz
Department of Mechanical Engineering, University of Malaya, Kuala Lumpur 50603, Malaysia.
Mukesh Kumar
Quantum University, Dehradun Highway, Mandawar, Roorkee, Uttarakhand 247167, India
Dr. Seema Tiwari
Associate Professor Applied Science & General Engineering Army Institute of Technology, Pune.
Mr. John Abbas
IBIS Group, Lucknow, UP, India.
Prof. (Dr.) B. K. Sarkar (Patent Guru/ Expert)
MES, Geh Research, India, USA, Singapore, Japan, Austraila, Germany.
Prof. (Dr.) Reena Singh,
MES's, Pillai HOC College of Engineering &Technology, Rasayani Taluka Panvel, Dist, Navi Mumbai, Maharashtra 410207, India

ABSTRACT

As hydrogen fuel cell vehicles move from manifestation to commercialization, the users expect safe, convenient and customer-friendly fuelling. Hydrogen quality affects fuel cell stack performance and lifetime, as well as other factors such as valve operation. In this paper, previous researcher's development on hydrogen as a possible major fuel of the future has been studied thoroughly. Hydrogen is one of the energy carriers which can replace fossil fuel and can be used as fuel in an internal combustion engines and as a fuel cell in vehicles. To use hydrogen as a fuel of internal combustion engine, engine design should be considered for avoiding abnormal combustion. As a result it can improve engine efficiency, power output and reduce NOx emissions. The emission of fuel cell is low as compared to conventional vehicles but as penalty, fuel cell vehicles need additional space and weight to install the battery and storage tank, thus increases it production cost. The production of hydrogen can be 'carbon-free' only if it is generated by employing genuinely carbon-free renewable energy sources. The acceptability of hydrogen technology depends on the knowledge and awareness of the hydrogen benefits towards environment and human life. Recent study shows that people still do not have the sufficient information of hydrogen.

Keywords: *Hydrogen, Vehicle fuel, Engine, Emissions, Combustion*

INTRODUCTION

The main sources at present, to satisfy world's energy demand, are mainly fossil fuels, which are going to be depleted very fast. Fossil fuel resources are now clearly run through and their prices have become unstable presently. That is due to, first, influential economic acceleration mostly in China and India and, second, by economic recession. In pursuit of energy security, the challenges of controlling prices and the uncertain reserves are strong incentives [1]. Significant environmental and societal problems, such as global warming and local pollution are directly associated with excessive usage of fossil fuels. Such problems strongly stimulate the research, development and demonstrations of clean energy resources, energy carriers, and in the case of transportation and power trains. In recent study, hydrogen is one of the energy carriers that can replace fossil fuels, but further research is needed to expose its advantages and disadvantages before this alternative fuel can be commercialised. Hydrogen is the cleanest fuel having a heating value three times higher than petroleum. However, being man- made fuel the hydrogen is not natural source of energy, therefore, it involves production cost, which is responsible for it is three times more cost than petroleum products. There are still problems in the realization of the renewed hydrogen from water, but the market supply and the cost of hydrogen do not constitute the bottleneck of hydrogen vehicles today although the hydrogen used presently may not be renewed. But, hydrogen's excellent characters, studying the availability of H_2 in internal combustion (IC) engines, and investi- gating the performance of hydrogen fuelled engines, become one of the utmost important research directions for researchers. That is why in this study, we are going to review previous developments and studies that have been done by other researchers on hydrogen as a possible major fuel of the future, used as in an internal combustion engines and as a fuel cell in vehicles. The aim of this study is to review hydrogen as a fuel for internal combus- tion engines for the vehicle propulsion in terms of advantages, disadvantages and fundamentals of hydrogen engines. Whereas for vehicle fuel cells the study focuses on performances, cost, infrastructure, type of storage and type of productions in hydrogen [1–3].

HYDROGEN AS A FUEL IN INTERNAL COMBUSTION ENGINES

In the following sub-sections several aspects that are related to the use of hydrogen as a fuel in internal combustion engine will be discussed further [4]. The discussion includes properties of combustive hydrogen, abnormal combustion in hydrogen engine, engine components, thermal efficiency, emission production, power output, emissions and cost, hydrogen production plant, people acceptability of hydrogen fuelling station and life cycle of hydrogen [5,6].

Engine Concept

Hydrogen can be used in SI engine by three methods [7]:

1. By manifold induction

Cold hydrogen is introduced through a valve controlled passage into the manifold.

2. By direct introduction of hydrogen into the cylinder

Hydrogen is stored in the liquid form, in a cryogenic cylinder. A pump sends this liquid through a small heat exchanger where it is converted into cold hydrogen gas. The metering of hydrogen is also done in this unit. The cold hydrogen helps to prevent pre-ignition and also reduces NO_x formation. The arrangement of liquid hydrogen storage and details of hydrogen induction into the SI engine cylinder can be seen in Figs. 1 and 2, respectively [8,9].

3. By supplementing gasoline

Hydrogen can also be used as an add-on fuel to gasoline in SI engine. In this system, hydrogen is inducted along with gasoline, compressed and ignited by a spark.

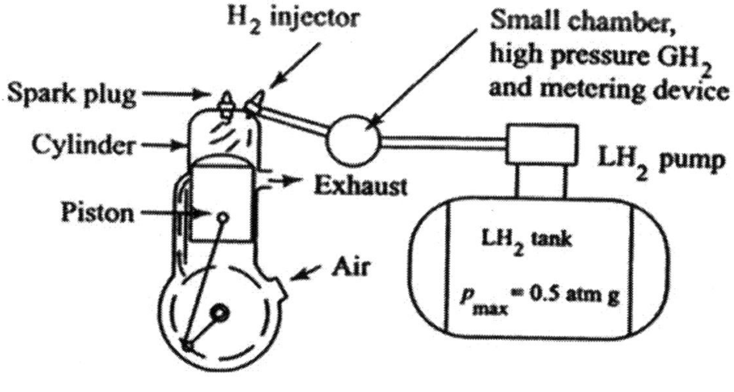

Figure 1: Liquid hydrogen storage and gaseous hydrogen injection [7].

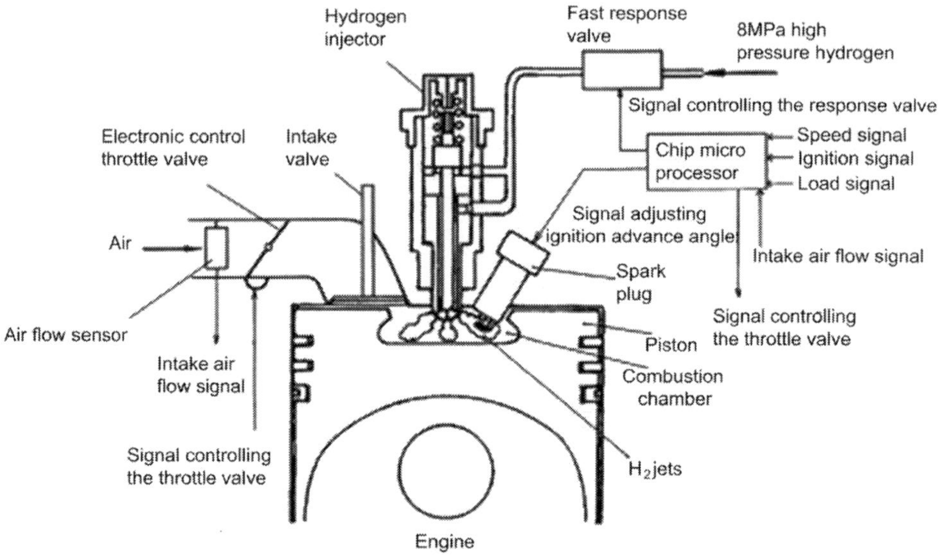

Figure 2: Hydrogen induction in spark-ignition engine [10].

Combustive Properties of Hydrogen

A brief summary of previous literatures are reviewed and discussed based on fundamentals of hydrogen combustion, Volumetric energy content flammability, ignition energy and octane number. Details of these characteristics to hydrogen engines based on recent studies as well as on-going efforts in the development of H_2ICEs and H_2ICE vehicles will be discussed later [11].

Some properties of hydrogen are listed in Table 1 in compar- ison with iso-octane and methane, which are representing as the natural gas and gasoline, respectively [12]. Table 2 shows the mixture properties of hydrogen–air when operated at lean and stoichiometric mixture in comparison with iso-octane–air and methane–air at stoichiometric mixture [13].

Flammability Limit

Flammability limit gives the proportion of combustible gases in a mixture; between these limits, this mixture is flammable. From Table 1 it is seen that the flammability of hydrogen in air (mixture) is at 4–75% which gives hydrogen wide range of flammability as compared to other fuels [14]. It is clear that 4% of hydrogen in air is still flammable but non-coherently and burns incompletely. The 4% value relates to

Table 1: Hydrogen properties compared with methane and iso-octane properties. Data given at 300 K and 1 atm, taken from [12].

Property	Hydrogen	Methane	Iso-octane
Molecular weight (g/mol)	2.016	16.043	114.236
Density (kg/m3)	0.08	0.65	692
Mass diffusivity in air (cm^2/s)	0.61	0.16	~0.07
Minimum ignition energy (mJ)	0.02	0.28	0.28
Minimum quenching distance (mm)	0.64	2.03	3.5
Flammability limits in air (vol%)	4.75	5-15	1.1–6
Flammability limits (λ)	10–0.14	2–0.6	1.51–0.26
Flammability limits (ψ)	0.1–7.1	0.5–1.67	0.66–3.85
Lower heating value (MJ/kg)	120	50	44.3
Auto-ignition temperature in air (K)	858	723	550
Flame velocity (ms^{-1})	1.85	0.38	0.37–0.43
Higher heating value (MJ/kg)	142	55.5	47.8
Stoichiometric air-to-fuel ratio (kg/kg)	34.2	17.1	15
Stoichiometric air-to-fuel ratio (kmol/kmol)	2.387	9.547	59.666

Table 2: Mixture properties of hydrogen–air, methane–air, and iso-octane–air. Data given at 300 K and 1 atm (with the exception of the laminar burning velocity, given at 360 K and 1 atm) [12].

Property	H_2–air $\lambda=1$ $\varphi=1$	H_2–air $\lambda=4$ $\varphi=0.25$	CH_4–air $\lambda=1$ $\varphi=1$	C_8H_{18}–air $\lambda=1$ $\varphi=1$
Volume fraction fuel (%)	29.5	9.5	9.5	1.65
Mixture density (kg/m^3)	0.85	1.068	1.123	1.229
Kinematic viscosity (mm^2/s)	21.6	17.4	16	15.2
Auto-ignition temperature (K)	858	>858	813	690
Adiabatic flame temperature (K)	2390	1061	2226	2276
Thermal conductivity (10^{-2} W/mK)	4.97	3.17	2.42	2.36
Thermal diffusivity (mm^2/s)	42.1	26.8	20.1	18.3
Ratio of specific heats	1.401	1.4	1.354	1.389
Speed of sound (m/s)	408.6	364.3	353.9	334
Air-to-fuel ratio (kg/kg)	34.2	136.6	17.1	15.1
Mole ratio before/after combustion	0.86	0.95	1.01	1.07
Laminar burning velocity, ~360 K (cm/s)	290	12	48	45
Gravimetric energy content (kJ/kg)	3758	959	3028	3013
Volumetric energy content (kJ/m^3)	3189	1024	3041	3704

configuration of one particular experiment. Therefore, the limit may vary being below 4% or above, (depending on condition), in real-world situations. For safety considerations this limit is important where it is less important for engine combustion [15]. Wide ranges of mixture of hydrogen permit extremely lean or rich mixture that combust with air. This makes the hydrogen engine operate at lean mixture resulting in greater fuel economy and more complete combustion reaction [16]. Final combustion temperature will also generally lower due to lower laminar burning velocity as can be seen in Table 2.The burning velocity for hydrogen engine that operates at lean mixture is rapidly lowered as compared to hydrogen engine that runs on stoichiometric mixture which is 12 cm/s (at φ=0.25) and 290 cm/s (at φ=1). This absolutely will reduce amount of pollutants such as NO_x [17].

Minimum ignition energy

Minimum ignition energy is the minimum amount of energy required to ignite a combustible vapour or gas mixture. At atmospheric conditions, the minimum ignition energy of a hydrogen–air mixture is an order of magnitude lower than for the mixtures of iso-octane–air and methane–air. For hydrogen concentrations of 22–26% only 0.017 MJ is obtained. Normally, capacitive spark discharge is used to measure minimum ignition energy, and thus is dependent on the spark gap [12].The values quoted in Table 1 are for a 0.5 mm gap. The minimum ignition energy can increase about 0.05 MJ and more or less constant for hydrogen concentrations between 10% and 50% when using a gap of 2 mm. The benefits for having minimum ignition energy to enable hydrogen engine to ignite lean mixture and ensure prompt ignition [18]. But having minimum ignition energy will increase possibility for hydrogen air mixture in the combustion chamber to be ignited by any other source (hot spot) rather than spark plug [13].

Small quenching distance

As compared to gasoline and other fuels, hydrogen has small quenching distance. In Table 1 the quenching distance for hydro- gen is about 0.64 mm compare to methane which is 2.03 mm and Iso-octane 3.5 mm. This parameter measures how close hydrogen flames can travel closer to the cylinder wall before they extin- guish. The smaller the distance, more difficult to quench the flame and this will increase the tendency for backfire. Experimentally, from the relation between minimum ignitions energy and the spark gap size quenching distance can be derived or can be measured directly [12].

High auto-ignition temperature

Referring to Table 1, taken from [18] hydrogen has relatively high auto-ignition temperature as compared to methane and iso- octane which is 858 K. This high auto-ignition is important parameter to determine engine compression ratio, since during compression, the temperature rise is pertained to the compres- sion ratio when considering Otto cycle [19] as shown in the Eq. (1) below.

$$T_2 = T_1 \left(r^c\right)^{k-1} \quad \ldots(1)$$

From this equation it can be seen that compression ratio is dependent on T_2 which is temperature during compression. This T_2 is limited by auto-ignition temperature to prevent fuel air mixture to auto ignite before the spark, given from spark plug. Higher auto-ignition temperature will increase T_2 and simulta- neously increase compression ratio. As relating to thermal effi- ciency of the system, higher compression ratio is important [20].

High flame speed, high diffusivity and low density

At stoichiometric ratios, hydrogen acquires high flame speed as shown in Table 1, which is about 1.85 ms^{-1} compared to methane and iso-octane which is 0.38 ms^{-1} and 0.37–0.43 ms^{-1}, respectively. Having high flame speed, hydrogen engines can more be similar to the thermodynamically ideal engine cycle. However, the flame velocity goes to decreases significantly at leaner mixture, [18]. Hydrogen also possesses remarkably high diffusivity, which is its capability to disperse in air more than methane and iso-octane. This shows that hydrogen can form uniform mixture of fuel and air, and if hydrogen leaks, it will disperse rapidly and leaking hydrogen is not a pollutant to the environment. Low density of hydrogen will result in two pro- blems of IC engine. Large volume needs to store more hydrogen to provide sufficient driving range and reduce power output due to low energy density [21].

Abnormal Combustion

The main problem to use hydrogen as a fuel in internal combustion engine, is to control the undesired combustion phenomena due to low ignition energy, wide flammability range and rapid combustion speed of hydrogen that causes mixture of hydrogen and air to combust easily [22]. In this section the main abnormal combustion in hydrogen engine which are pre-ignition, backfire, and knock in terms of cause and method to avoid will be discussed.

Pre-ignition

Pre-ignition is one of the undesired combustion that needs to be avoided in hydrogen engine. During the engine compression stroke, these abnormal combustion events occur inside the combustion chamber, with actual start of combustion prior to spark timing [23]. Pre-ignition event will advance the start of combustion and produce an increased chemical heat-release rate. In turn, the increased heat-release rate results in a rapid pressure rise, higher peak cylinder pressure, acoustic oscillations and higher heat rejection that leads to rise in-cylinder surface tem- perature. The start of combustion can further be advanced by latter effect, which in turn can be led to runaway effect, and will cause the engine failure if unchecked [24]. It is observed from Fig. 3, that as the stoichiometric condition ($\varphi = 1$) is approached from the lean side ($\varphi < 1$), the minimum ignition energy for hydrogen is a strongly decreasing function of the equivalence ratio with the minimum at $\varphi \approx 1$ This trend shows that it isextremely difficult to operate an H_2ICE at or near the stoichio- metric condition in the absence of frequent pre-ignition events.

Therefore, the maximum j and, consequently, peak power output can be limited by the pre-ignition limit for practical application. Stockhausen et al. [25] report a pre-ignition limit of $\varphi \approx 0.6$ for a 4-cylinder 2.0-*l* engine at an engine speed of 5000 rpm. Although the pre-ignition limit is engine specific, the consistent trends with variations in engine properties and operational conditions have been found: the pre-ignition limited φ decrease monotonically with increased compression ratio (CR) [25,26] and increase mixture temperature [25]. An effect on engine speed has also been shown [26] but due to the coupled effect of residual mass fraction the trend is more complicated.

From above description, it is clearly seen that pre-ignition limit will border on the peak power output of hydrogen engine and this will decrease the performance of H_2ICE poweredvehicle in comparison to its gasoline equivalent [27]. Therefore, determining the mechanism of pre-ignition, practical operational limits, and control strategies has been a primary focus of many research studies. Unfortunately, there are still no guaranteed preventive steps, but identification of pre-ignition source has provided the necessary minimizing steps.

Source of pre-ignition;

- Hot spark plugs or spark plug electrodes.
- Hot exhaust valves or other hot spots in the combustion chamber.

- Residual gas or remaining hot oil particles from previous combustion events.
- Combustion in crevice volumes [18].

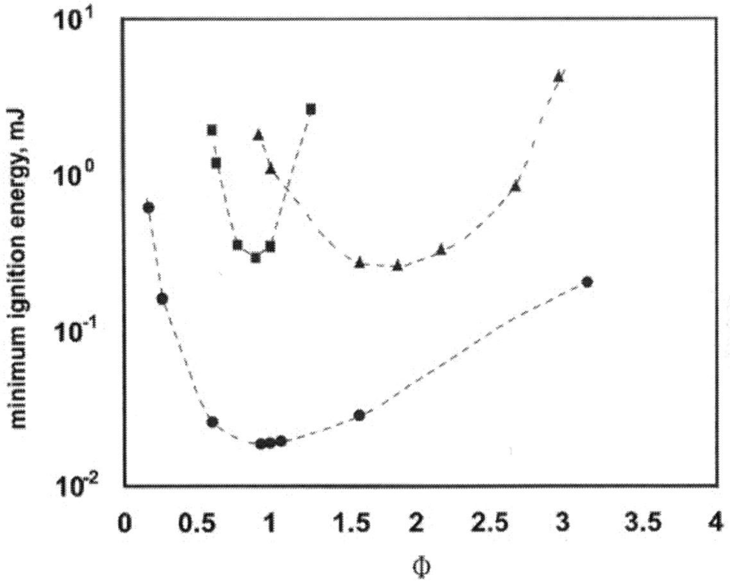

Figure 3: Minimum ignition energies of (●) hydrogen–air, (■) methane–air and (▲) heptane–air mixture in relation to at atmospheric pressure [18].

As the minimum ignition energy is dependent on the equiva- lence ratio, the pre-ignition becomes more pronounced when the hydrogen–air mixture approach stoichiometric levels. At increased engine speed and load, operating conditions will also be prone to the occurrence of pre-ignition due to higher gas and components temperature [12].

Several steps have been taken to minimize the source of pre- ignition which are [12]:

- Proper design of spark plug.
- Ignition system design with low residual charge. Specific design of crankcase ventilation.
- Sodium-filled exhaust valve [19].
- Optimized design of the engine cooling passage to avoid hot spot.
- With the use of hydrogen direct injection systems.
- Variable valve timing for effective scavenging of exhaust residuals [18].

Kondo et al. [18] used an ignition system specification, designed to prevent residual energy and a water-cooled spark plug. From the test conducted, variance of equivalence ratio based on advanced control strategies is shown in Table 3.

Backfire

Backfire is one of the main problems to run a hydrogen fuelled engine. Backfire or flashback is the uncontrolled combustion of fresh hydrogen–air mixture during the intake stroke in the combustion chamber and/ or the intake manifold. The fresh hydrogen–air mixture is aspirated into the combustion chamber with the opening of the intake valves. Backfiring is caused when combustion chamber hot sopts, hot residue gas or remaining charge in the ignition system ignite the fresh charge as hydrogen has low ignition temperature [28]. This abnormal combustion occurs due to the same concept of pre-ignition. The difference is the timing at which the anomaly occurs. In pre-ignition, the uncontrolled combustion happens during compression stroke when the intake and exhaust valves close before spark plug fires in cylinder [22,27], while backfire occurs during intake stroke when the intake valve is opened. The backfire initiates from the pre-ignition during the compression stroke, and then proceeds to the ignition of the intake mixture [29].

Table 3: Effect of advance control strategies to the equivalence ratio limit to the pre- ignition occurrence [18].

Equivalence ratio	Advance control strategies
$\varphi \approx 0.35$	Without any advanced control
$\varphi \approx 0.6$	Elimination of residual energy in the ignition system
$\varphi \approx 0.8$	With addition of the water-cooled spark plug

Effect of backfire resulting in combustion and pressure rise in the intake manifold, is clearly audible as well as can also damage or destroy the intake system. When mixture approaches stoichiometry, the occurrence of backfire is more likely due to the low ignition energy, and when using PFI-H_2ICE, as the hydrogen is injected before the intake valve opens, to mix with air in the intake manifold before entering combustion chamber. While in DI-H_2ICE, the occurrence of backfire can be neglected as hydrogen injection starts after the intake valve closes difference with external mixture formation concept [12]. Recently, many works have been carried out on optimizing the intake design and injection strategies to avoid backfiring. Conse- quently, the measures those help in avoiding pre-ignition also reduce the risk of backfiring. Some of the strategies that are used to avoid backfiring:

- Injection strategies that allow pure air to flow into the combustion chamber to cool potential hot spots before aspir- ating the fuel-air mixture.
- The possibility of backfiring mainly depends on the concentrations of H_2 residual at intake ports in a manifold injection H_2ICE, and the leaner the concentration of the residual, the lower the possibility of the backfire.
- Optimization of the fuel-injection strategy in combination with variable valve timing for both intake and exhaust valves allow operation of a port injected hydrogen engine at stoichio- metric mixtures over the entire speed range.

Knock

Knock, or spark knock [24] is defined as auto-ignition of the hydrogen–air end-gas ahead of the flame front that has originates from the spark. This follows a rapid release of the remaining energy generating high-amplitude pressure waves, mostly referred to as engine knock. Engine damage can be caused by the amplitude of the pressure waves of heavy engine knock due to increased mechanical and thermal stress [1]. The knocking tendency of an engine is dependent on the engine design along with the fuel-air mixture properties. The high auto-ignition temperature, finite ignition delay and the high flame velocity of hydrogen mean that knock, as defined is less likely for hydrogen relative to gasoline, and hence the higher research octane number (RON) for hydrogen (RON4120) in com- parison with gasoline [18].

Following are the effects of knock to engine operation [25]:

- Increased heat transfer to the cylinder wall.
- Excessively high cylinder pressure and temperature levels and increased emissions.
- Undesirable engine performance and the potential damage to engine components.

It is critically important not only to avoid knocking but also to know the limiting conditions for its incidence under any set of operating and designing conditions. Effective means for extending knock-free operation need to be developed. Several tests have been done to understand knock behaviour and detection in hydrogen engine. From [30] knock is influenced by parameters including the engine compression ratio, the type of fuel, ignition timing, and the fuel-air-dilution mixture. Several results are drawn based upon the observation and analysis in this work:

- It is rational to say that hydrogen knock can be treated similar to gasoline knock for practical engine applications. As a result, gasoline knock detection and potential engine control techniques can be extended for use with hydrogen without significant changes.
- The combustion knocks level probability distributions for both hydrogen and gasoline changes as the knock level increases. The skewness of the distribution reduces as the overall knock level increases.
- Changes required from gasoline combustion knock detection and control would result primarily from hydrogen's sensitivity to engine operating conditions including the engines compression ratio, stoichiometry, dilution levels, ignition timing and the differences in operating conditions for a hydrogen engine to obtain optimal performance.

Fig. 4 [26] shows the effect of compression ratio and equiva- lence ratio on the engine power, at optimum spark timing for best torque and 25 rps engine speed. The figure shows that the high useful compression ratio (HUCR), which gives the highest power, occurred at the compression ratio (CR) of 11:1. With further increase in compression ratio, the engine power decreases due to unstable combustion [23,31]. However, the loss of combustion control, which is pre-ignition, limits the maximum power output of a hydrogen engine. As referred to Fig. 4, it is observed that operating at lean or rich mixtures tends to decrease the engine power for all compression ratios. Air-to-fuel ratios rich of stoi- chiometric decrease engine power due to decreasing combustion efficiency. Air-to-fuel ratios lean of stoichiometric decrease engine power due to a reduction in the volumetric lower heating value of the intake mixture, despite increasing combustion efficiency. Fig. 5 [26] show the effect of engine speed on the engine power, at optimum spark timing for best torque and HUCR. It is clearly seen that as the engine power increases engine speed increases [32].

Avoiding abnormal combustion

It is an effective measure to limit maximum fuel-to-air equivalence ratio to avoid abnormal combustion in hydrogen operation. By operation, employing a lean-burn strategy, the excess air in lean operation acts as an inert gas and reduces combustion temperature effectively and components temperatures consequently.

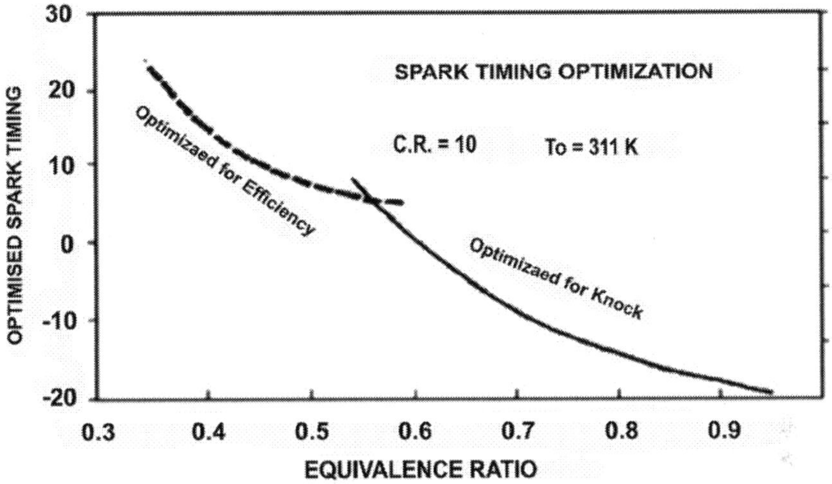

Figure 4: Typical variations in spark timing for optimum efficiency and avoidance of knock for lean mixture operation with hydrogen [33].

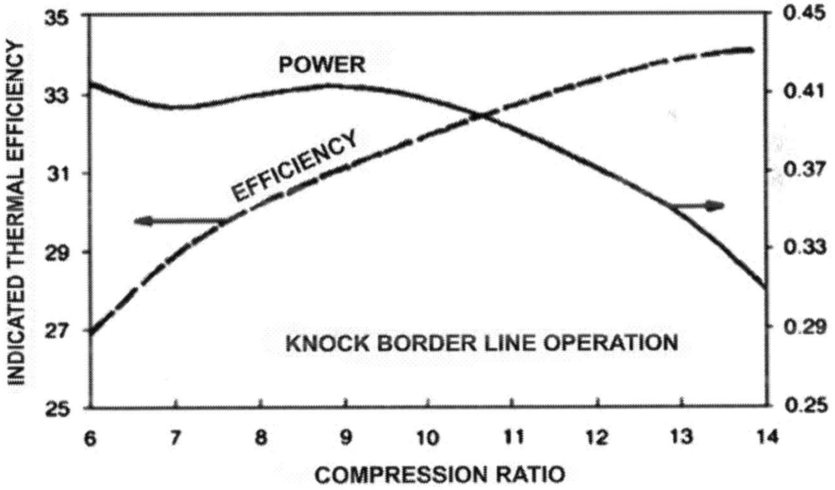

Figure 5: Typical variations in indicated power output and efficiency with changes in compression ratio when using optimum spark timing for borderline knock [33].

Although lean operation is very effective, it does limit the power output of hydrogen engine. Using thermal dilution technique, pre-ignition conditions can also be curbed, such as water injection or exhaust gas recirculation (EGR). A portion of the exhaust gases is re-circulated back into the intake manifold by EGR system. It helps to reduce the tem- perature of hot spots by introducing the exhaust gases, hence reducing the possibility of pre-ignition. Additionally, peak combustion temperature is reduced by recirculation of exhaust gases, which also reduces NO_x emissions. Typically, a 25% to 30% recirculation of exhaust gases is effective in elimination of back fire [34].

Injection of water is the other technique for thermally diluting the fuel mixture. Injecting water into the hydrogen stream prior to mixing with air has produced better results than injecting it into the hydrogen–air mixture within the in-take manifold. A potential problem of mixing of water with oil exists with this type of system, so care must be taken ensuring that seals do not leak.

Engine Components

Some features of engines designed for or converted to hydro- gen operation, will be discussed in this section. As discussed in the previous section, the occurrence of combustion anomalies, or more particularly, the desire to prevent it, has led to most of the countermeasures, which were put forwarded in the early work on H_2ICEs [35].

Spark plugs

To avoid spark plug electrode temperature that exceeds the auto-ignition limit and causing backfire, cold-rated spark plugs are recommended [36]. This cold spark plug can be used since there are no carbon deposits to burn off. Since spark plugs with platinum electrodes can be catalyst to hydrogen oxidation, there- fore these are to be avoided.

Injection systems

In hydrogen engine, there are two types of injection systems, which can be used, one is port fuel injection (PFI) and other is direct injection (DI). In PFI-H_2ICE, time injection is a prerequisite as what has already been discussed previously that the main problem in PFI-H_2ICE is to avoid backfire. Therefore, PFI needs the programming of the injection timing such that an air cooling period is created in the initial phase of the intake stroke, and the end of injection is such that all hydrogen is inducted, leaving no hydrogen in the manifold when the intake valve closes. The advantage of using PFI system is the pressure tank for injector, which is lower as compared to DI system.

In DI-H_2ICE, hydrogen is injected directly into the combustion chamber during compression stroke. Hydrogen injection at compression stroke prevents knock and gives an increase in thermal efficiency and maximum output power [37].

Hot spots

Minimizing the hot spot in combustion chamber of hydrogen engine is important to avoid abnormal combustion which is the major problem in burning hydrogen well because it will reduce power output and engine efficiency. Hot spot can act like ignition source as hydrogen needs minimum ignition energy to be ignited. There are several steps to minimize hot spots in combustion chamber as few are given as following [12,38]:

- Using cooled exhaust valve; multi-valve engine leads to further bringing the temperature of exhaust valve down.
- Additional engine coolant passages around valves and other area with high thermal loads.
- Adequate scavenging to decrease residual gas temperature.

Piston rings and crevice volumes

Previously, many experiments [12] have been conducted to eliminate all hot spots (e.g., careful cleaning of the engine, enhanced oil control or even, scavenging of the residual gases, cold spark plugs, cooled exhaust valves, etc.), where, backfire and uncontrolled spark-induced ignition still occurred. On the other hand, hydrogen engines have been demonstrated, run on stoi- chiometric mixture without any occurrence of backfire, by careful selection of crevice volumes and piston rings, without any need for timed injection or cooled exhaust valve. Therefore, it needs careful selection of piston rings and crevice volumes in order to prevent hydrogen flame from propagating into the top land [39].

Lubrication

Lubrication is an important aspect that needs to be considered when switching over to hydrogen as fuel in internal combustion engine. During engine operation, blow will always occur due to the rapid pressure rise and the low density of hydrogen gas. The exhaust gases, entering crankcase can condense, when there is no provision of proper ventilation. Water mixing into the crankcase oil (lubrication oil) reduces its lubrication ability and as a result, there occurs a higher degree of engine wear [12]. Measurement of the composition of the gases in the crankcase at Ghent University [40] showed a very high percentage of hydrogen arising from blow by. The investigation of the lubricating oil is carried out and is compared to that of the unused oil. The oil properties severely changed with a strong decrease in the lubricating qualities [41].

Engine lubrication oil having compatibility with increased water concentration in the crankcase, has to be chosen. Engine specific oil, which is developed for hydrogen engines, is probably the best solution but currently is not available [42].

Crankcase ventilation

Using hydrogen as a fuel for spark-ignited internal combustion engines, especial attention has to be given to the crankcase ventilation as compared to gasoline engines. Carbon based deposits from the engine's lubricating oil, in the combustion chamber, on the top of the piston, in the ring grooves, and in the cylinder's squish areas are potential hot spots waiting to happen. Blow by effect can cause unburned hydrogen entering the crankcase and at certain concentration, hydrogen can combust in the crankcase due to lower energy ignition and wide flammability limits. Hydrogen should be prevented from accumulating through ventilation [12,43].

2.4.7. Compression ratio

It is the similar choice of the optimal compression ratio to that for any fuel; for increasing engine efficiency it should be chosen as high as possible, with the limit given by increased heat losses or the occurrence of abnormal combustion (in the case of hydrogen, primarily surface ignition). The choice may be depen- dent on the application, as the optimum compression ratio for highest engine efficiency might be different from the optimum compression ratio for highest power output. Compression ratios used in H_2ICEs range from 7.5:1 to 14.5:1 [12,44].

In-cylinder turbulence

Low turbulence combustion chamber can be used due to high flame speeds of hydrogen. Low radial and tangential velocity components can be reduced by the use of disk-shaped combus- tion chamber (flat piston and chamber ceiling) can help produce and does not amplify inlet swirl during compression. This will be beneficial for engine efficiency by increase in the volumetric efficiency and decrease heat loses [42]. The overall trends are such that turbulence increases auto-ignition delay times and accordingly the ignition length and pressure further contribute to this delay [45].

Materials

Hydrogen effects on the mechanical properties of iron as embrittlement, such as decrease in true stress of fracture and ductility [46].

Types of hydrogen embrittlement of steels [12]:

- As concentrations of hydrogen occur on surface, the hydrogen reaction embrittlement is arisen, resulting in chemical reaction.
- Environmental embrittlement: in the atmosphere containing hydrogen, there happens adsorption of molecular hydrogen on the surface and its absorption within the lattice after dissociation into atomic form.
- It happens in the absence of hydrogenated atmosphere and due to the hydrogen that enters the lattice during processing or fabrication of steel.

Brass and copper alloys, aluminium and aluminium alloys, and copper beryllium are the materials which can be used for the applications of hydrogen. Nickel and high-nickel alloys, titanium and titanium alloys are very sensitive to hydrogen embrittlement. In the case of steel, for hydrogen embrittlement sensitivity, it depends on the exact chemical composition, heat or mechanical treatment, microstructure, impurities and strength [41].

Thermal Efficiency

The theoretical thermodynamic efficiency of an Otto cycle engine is based on the compression ratio of the engine as shown in Eq. (2)

$$\eta_{th} = 1 - \left(\frac{1}{r^c}\right)^{\gamma - 1} \qquad \ldots(2)$$

The higher compression ratio r^c and/or the specific heat ratio g, indicated the thermodynamic efficiency of the engine. Hydrogen ($\gamma = 1.4$) has much simpler molecular structure than gasoline and therefore its specific–heat ratio is higher than that of gasoline ($\gamma = 1.1$). As a result, theoretically, hydrogen engine can have higher thermal efficiency compared to gasoline engine [44]. The high RON and low-flammability limit of hydrogen provides the necessary elements to attain high thermal efficiencies in an internal combus- tion engine [18]. In DI-H2ICE, hydrogen injection at later stage of compression stroke can achieve the thermal efficiency higher than 38.9% and the brake mean effective pressure 0.95 MPa [37].

Thermodynamic analysis

Using test data from different operating modes, the engine efficiencies and the losses of the working cycle can be calculated. Fig. 6 shows the efficiencies and losses for gasoline and hydrogen operation with both port injection and direct injection.

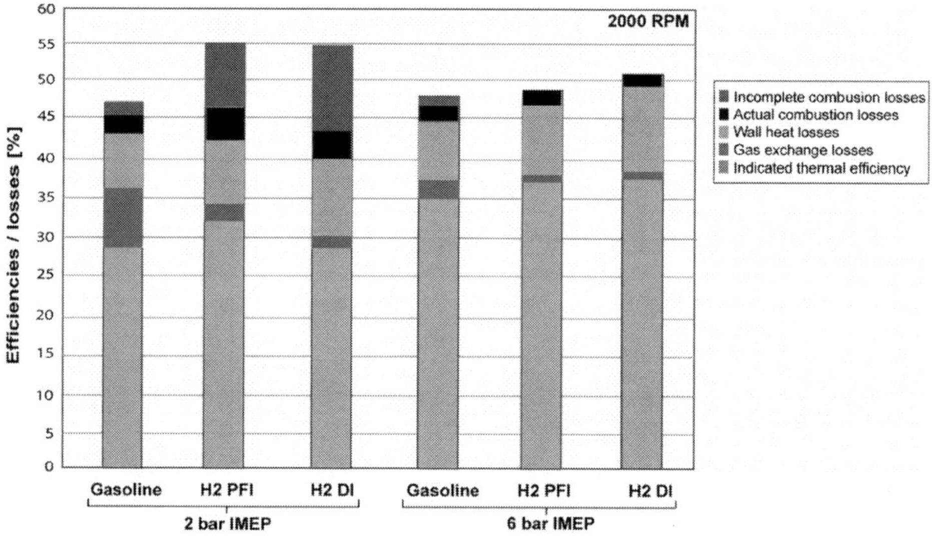

Figure 6: Analysis of losses compared to the theoretical engine cycle; gasoline versus hydrogen (PFI and DI), at two loads [12].

The data for all fuels were collected on a single-cylinder research engine at an engine speed of 2000 RPM and indicated mean effective pressure of 2 bar and 6 bar. Table 4 shows summary of the analysis [47].

Comparative combustion characteristics of gasoline and hydro- gen fuel, in internal combustion engine have been done [48]. The ability of elucidating the potential performance and efficiency of a hydrogen fuelled ICE compared to a gasoline fuelled ICE was achieved in the analysis of the comparative combustion character- istics of hydrogen and gasoline fuelled internal combustion engine. It was noted that the hydrogen fuelled ICE had a higher thermal efficiency of approximately 6.42% due to the reasons as; less heat rejection during the exhaust stroke, less blow down during the exhaust stroke, combustion taking place closer to TDC and combus- tion taking place in an closer to isochoric environment. Thus, it was closer to an actual Otto cycle [49]. An important conclusion is that improvement in H_2ICE efficiencies will require strategies to mini- mize heat transfer losses to the cylinder walls as higher combustion temperatures and shorter quenching distance associated with hydrogen combustion are believed to cause the greater convective heat transfer to the cylinder walls [42] Table 5.

Table 4: Result analysis of losses compared to the theoretical engine cycle at load 2 bar IMEP. Efficiency/losses Cause

Efficiency/losses	Cause
AT 2 bar IMEP	
Efficiency of the ideal engine in gasoline operation is lower than in H_2	Compression ratio and AF ratio higher in H_2 due to lean operation
Incomplete combustion losses	Due to extremely lean condition in H_2 operation
Actual combustion losses—in gasoline around 3% lower than H_2	Due to the lean combustion in H_2
Wall heat losses in H_2 operation higher than gasoline	Higher pressure levels in H_2 operation resulting from unthrottled operation
Wall heat losses in H_2 DI are higher than PFI.	Due to higher in-cylinder charge motion
Gas exchange losses in H_2 operation are only fraction compared to gasoline	Since the engine is operated unthrottled
Overall indicated thermal efficiency for H_2PFI is higher than gasoline & H_2DI	

Table 5: Result analysis of losses compared to the theoretical engine cycle at load 6 bar IMEP. Efficiency/losses Cause

Efficiency/losses	Cause
AT 6 bar IMEP	
Efficiency of the ideal engine in gasoline operation is lower than in H_2	Compression ratio and AF ratio higher in H_2 due to lean operation PFI lower than DI due to the air displacement effect
Incomplete combustion losses in gasoline is more than 1%, H_2PFI & H_2DI less than 0.5%	Very complete combustion in H_2 due to the fast flame speed & small quenching distance
Actual combustion losses—in gasoline around 2% and H_2 is lower	H_2: unthrottled & lean mixture, the combustion still faster than gasoline
Wall heat losses in H_2 operation higher than gasoline	Due to higher flames speeds and the smaller quenching distance
Wall heat losses in H_2DI operation higher than H_2PFI operation	Due to higher level of in-cylinder charge motion and turbulence caused by DI event.
Gas exchange losses in H_2 operation are lower compared to gasoline	Due to engine operated unthrottled.
Overall indicated thermal efficiency for H_2 more than 2.5% both with DI and PFI compare to gasoline.	

Emission Production

Because of the reasons that hydrogen can be produced form any kind of energy source and it is combusted without emitting carbon dioxide or soot, it is considered as an ideal alternative fuel to conventional hydrocarbon fuels [50]. The only potential emis- sions are the nitrogen oxides (NO_x) as pollutants from hydrogen combustion, hence it becomes crucial to minimize the (NO_x) emissions from the combustion of hydrogen. Eqs. (3) and (4) show the exhaust gas emission from hydrogen which is water and NO_x [38].

$$2H_2 + O_2 = 2H_2O \quad \ldots(3)$$

$$H_2 + O_2 + N_2 = H_2O + N_2 + NO_x \quad \ldots(4)$$

The formation of nitrogen oxides occurs, because the higher temperatures are generated within the combustion chamber during combustion. These higher temperatures cause some of the nitrogen and oxygen to combine, present in the air [51]. The technique of rich-lean combustion or staged combustion is used to reduce NO_x formation in continuous combustion burners such as gas turbines and boilers [52]. Where, water injection in the compression ignition engine helps to control combustion tem- perature and pressure. Hence, it is beneficial in controlling unwanted emissions. Many researchers have demonstrated the effect with conventional hydrocarbon fuels [53].

The amount of NO_x formed depends on;

- The air/fuel ratio.
- The engine compression ratio.
- The engine speed.
- The ignition timing.
- Thermal dilution is utilized or not [54].

Power Output

Volumetric efficiency, fuel energy density and pre-ignition primarily determine the H_2ICE peak power output. The volu- metric efficiency has been proved to be the limiting factor for determining the peak power output for most of the practical applications. The displacement of intake air by the large volume of hydrogen in the intake mixture is the reason for PFI-H_2ICEs to inherently fusser from volumetric efficiency. For example, about 30% of hydrogen is possessed by mixture of hydrogen and air by volume, whereas a 2% gasoline is possessed by stoichiometric mixture of fully vaporized gasoline and air by volume. The higher energy content of hydrogen partially offsets the corresponding power density loss. The stoichiometric heat of combustion per standard kg of air is 3.37 MJ and 2.83 MJ for hydrogen and gasoline, respectively. It follow that approximately 83% is the maximum power density of a pre-mixed or PFI-H_2CE, relative to the power density of the gasoline operated identical engine [18]. For applications where peak power output is limited by pre- ignition, H_2ICE power densities, relative to gasoline operation, can significantly be below 83%. For direct injection systems, which mix the fuel with the air after the intake valve closes (and thus the combustion chamber has 100% air), the maximum output of the engine can be approximately 15% higher than that for gasoline engines. Therefore, depending on how the fuel is metered, the maximum output for a hydrogen engine can be either 15% higher or 15% less than that of gasoline if a stoichiometric air/fuel ratio is used [55].

However, at a stoichiometric air/fuel ratio, the combustion temperature is very high and as a result it will form a large amount of nitrogen oxides (NO_x), which is a criteria pollutant. Since one of the reasons for using hydrogen is low exhaust emissions, therefore hydrogen engines are not normally designed to run at a stoichiometric air/fuel ratio. At this air/fuel ratio, the formation of NO_x is reduced to near zero. Unfortunately, this also reduces the power out-put. To make up the power loss, hydrogen engines are usually larger than gasoline engines, and/or are equipped with turbochargers or superchargers [47].

Emissions and Cost

In future, main goal of development of energy is to get the best efficiency with the affordable cost. In order to achieve this goal, there are a few things, which should be considered. In terms of vehicles the production cost, the fuel cost, and the environmental impacts should be considered [38]. In the previous study, a comparison between conventional, hybrid, electric or battery, hybrid fuel cell and battery and fuel cell vehicle has been investigated in terms of fuel cost, production cost and air pollu- tion emission [47,56].

From the first study, it summarizes that Tables 6 and 7, the (1–3) represent types of considerations, and it is as:(1) electricity is produced from renewable energy sources including nuclear energy; (2) 50% of the electricity is produced from renewable energy sources and 50% from natural gas with an efficiency of 40%; (3) all electricity is produced from natural gas with an efficiency of 40% [57]. According to those results, hybrid and electric cars are competitive if nuclear and renewable energies account for about 50% of the energy to generate electricity. If fossil fuels (natural gas) are used for more than 50% of the energy to generate electricity, the hybrid car has significant advantages over the other three [58]. The fuel cell shows quite low efficiency due to the production of hydrogen generated high air pollution and also greenhouse gases emission that reduce its efficiency. The air pollution is analysed using the curb weight of the vehicle and that increases the fuel cell air pollution emission [38,56].

Table 6: Normalized and environmental indicators for four types of car [56].

Car	Normalized indicators					General indicator	Normalized general indicator
	Car	Range	Fuel cost	Greenhouse gas emission	Air pollution emission		
1[a]							
Conventional	1	0.581	0.307	0.108	0.126	0.00243	0.0651
Hybrid	0.733	1	0.528	0.174	0.205	0.0138	0.37
Electric	0.212	0.177	1	1	1	0.0374	1
Fuel cell	0.154	0.382	0.532	0.163	0.247	0.00126	0.0336
2							
Conventional	1	0.581	0.307	0.336	0.436	0.0261	0.176
Hybrid	0.733	1	0.528	0.541	0.708	0.148	1
Electric	0.216	0.177	1	1	1	0.0374	0.252
Fuel cell	0.154	0.382	0.532	0.488	0.807	0.0123	0.0832
3							
Conventional	1	0.581	0.307	0.599	0.628	0.067	0.197
Hybrid	0.733	1	0.528	0.911	0.967	0.341	0.341
Electric	0.212	0.177	1	1	0.824	0.0308	0.0908
Fuel cell	0.154	0.382	0.532	0.794	1	0.0248	0.0728

Table 7: Greenhouse gas and air pollution emissions related to the fuel utilization stage and total environmental impact for different types of cars [56].

Car type	Fuel utilization stage		General indicator	
	GHG emissions per 100 km of vehicle travel (kg per 100 km)	AP emissions per 100 km of vehicle travel (kg per 100 km)	GHG emissions per 100 km of vehicle travel[a] (kg per 100 km)	AP emissions per 100 km of vehicle travel (kg per 100 km)
Conventional	19.9	0.0564	21.4	0.06
Hybrid	11.6	0.0328	13.3	0.037
1[b]				
Electric	0.343	0.00131	2.31	0.00756
Fuel cell	10.2	0.0129	14.2	0.0306
2				
Electric	5.21	0.0199	7.18	0.0262
Fuel cell	10.6	0.0147	14.7	0.0324
3				
Electric	10.1	0.0385	12	0.0448
Fuel cell	11.1	0.0165	15.2	0.0342

In other study the comparison is made between internal combustion engines (ICE), Fuel cell vehicle (FCEV), battery vehicle (BEV), and Fuel Cell & battery Hybrid vehicle (FCHEV). This study considers the cost of power train of the vehicle and also the fuel costs. It is estimated in terms of optimistic, pessimistic and average cost of the estimated power train costs and fuels in year 2030 [59]. From the study results, it shows that if the cost predictions for fuel cell, battery, hydrogen and electricity are correct, then for this scenario both the FCHEV and BEV options are the cheapest by 2030 in terms of lifecycle costs. The results also show that by 2030 the FCEV costs will have approached parity with the internal combustion engine costs as per Tables 8 and 9 [60].

Table 8: Summary of the capital cost input data [60].

	2010	2030 optimistic	2030 pessimistic	2030 average
Power train cost				
20 kW fuel cell	10,000	700	1,500	1,100
80 kW fuel cell	43,700	4900	10,030	7,465
6 kW h battery pack	6,000	1200	1,800	1,500
25 kW h battery pack	25,000	5000	7,500	6,250
Electric motor and controller	1,700	1200	2,030	1,615
Hydrogen storage	2,000	900	2,000	1,450
Conventional (ICE)	2,200	2400	2,530	2,465
Total cost				
ICE	2,200	2400	2,530	2,465
FCEV	47,400	7000	14,060	10,530
BEV	26,700	6200	9,530	7,865
FCHEV	19,700	4000	7,330	5,665

Table 9: Summary of the running cost input data [60].

Fuel cost	2010 (GJ)	2030 Optimistic (G/J)	2030 Pessimistic (G/J)	2030 Average (G/J)	Miles (G/J)	Typical units
Gasoline	12.7	19	38	28.5	253	40 mpg
Hydrogen	42	14	56	35	506	72 miles/kg
Electric	36	27	45	36	1013	3.6 mileskW/h

In term of capital cost, in 2010 FCEV and BEV and FCHEV are far more costly than conventional ICE power trains. But in year 2030 capital cost could drop significantly, with the FCHEV is the lowest followed by BEV and FCEV. In terms of fuel cost per miles, electric vehicles achieve the highest miles per GJ than hydrogen and gasoline vehicles. In 2030, BEVs and FCHEVs are relatively insensitive to fuel (electricity) cost changes, whereas FCEVs and ICEs exhibit marked sensitivity to hydrogen and gasoline costs, respectively. This is partly due to the differing power train efficiencies and lastly the total lifecycle costs over 100,000 miles. FCHEVs appear to be slightly cheaper than BEVs but exhibit a wider overall sensitivity to combine (capital and running) costs. Both ICEs and FCEVs have much greater lifecycle costs than FCHEVs and BEVs, around 1.75 times higher [56,60].

It is the source of electricity on which economics and environmental impact associated with use of an electric car are dependent substantially. It will be advantageous for electric car to the hybrid vehicle, if electricity comes from renewable energy sources. It will be competitive for electric cars only if the electricity is generated on-board, when the electricity comes from fossil fuels. The initial cost for fuel cell and battery is higher compared to conventional but throughout the year of 2030 the price of both battery and also fuel cell is almost competitive with internal combustion engine, it is the impact of the improvement and availability increases throughout the year [56,60].

The summary of fuel cost and emissions is shown in Table 7. From table we can see that the most efficient and less emission car is the hybrid car. Hydrogen generates high cost in terms of production cost but it generates the lowest air pollution emis- sions and high in greenhouse gases compared to electric and hybrid vehicle [56,60] Tables 10 and 11.

Table 10: Summary of production cost, fuel cost, and emissions [56,60].

	Production average cost in 2030 (RM)	Fuel costs (G/J)	Greenhouse gas emissions (kg/100 km)	Air pollution emission (kg/100 km)
Gasoline	2,465	28.5	21.4	0.0600
Hydrogen	10,530	35	15.2	0.0342
Electric or Battery	7,865	36	12.0	0.0448
Hybrid	5,665	–	13.3	0.0370

Hydrogen Production Plant

The main challenge to develop hydrogen fuel cell vehicle is the infrastructure. In recent study in china the main technologies of producing hydrogen are natural gas steam reforming (NGSR), coal gasification, and water electrolysis. As for storage is concerned there are three ways in which it can be stored as; hydrogen gas, liquid hydrogen and hydride [61]. The development is considering the situations of resources, environment, energy supply and technical economy in China [41,62]. The problem is studied in view of "time" and "space". In terms of time, Coal dominates the energy structure in China; the infrastructure should change with the energy structure. Coal generates more CO emissions but produces low hydrogen. But the change of energy will reduce the usage of coal hence the usage and production of hydrogen via other primary source will increases. The technology progress will change the hydrogen infrastructure. In terms of space, the hydrogen infrastructure will probably be different in different region in China. In different region, the electricity is generated using different methods [63].

From Fig. 7, it is seen that plans 5 and 6 show the highest energy efficiency and plans 9 and 10 show the lowest energy efficiency. The conclusion for production is arranged from the best to the worst is coal gasification, NGSR, methanol reforming on-board and water electrolysis. And as for the storing and transporting methods, the ranking is done as hydrogen gas by pipeline, hydrogen gas by cylinder, liquid hydrogen, and Hydride [6,63].

Table 11: Summary of hydrogen plan [63,65].

Plan no	Production system	Transportation subsystem	Refueling subsystem	Utilization subsystem	Efficiency, environmental & ecoomic performance
1	Central factory: NGSR	Hydrogen gas cylinder by truck	Hydrogen gas cylinder	Hydrogen gas	• Low emission of air pollutant
2	Central factory: NGSR	Hydrogen gas by pipeline	Hydrogen gas tank	Hydrogen gas	• Low emission of air pollutant
3	Central factory: NGSR	Liquid hydrogen tank by truck	Liquid hydrogen tank	Liquid hydrogen	• Low emission of air pollutant
4	Central factory: NGSR	Hydride cylinder by truck	Hydride cylinder	Hydride	• Low emission of air pollutant
5	Central factory: coal Gasification	Hydrogen gas by truck	Hydrogen gas cylinder	Hydrogen gas	• High energy efficiency • 3rd lowest emission of air pollutant • 2nd lowest cost
6	Central factory: coal Gasification	Hydrogen gas by pipeline	Hydrogen gas tank	Hydrogen gas	• High energy efficiency • 3rd lowest emission of air pollutant • 2nd lowest in costs
7	Central factory: coal Gasification	Liquid hydrogen tank by truck	Liquid hydrogen tank	Liquid hydrogen	• 3rd lowest emission of air pollutant • 2nd lowest in costs
8	Central factory: coal Gasification	Hydride cylinder by truck	Hydride cylinder	Hydride	• 3rd lowest emission of air pollutant • 2nd lowest in costs
9	Refueling stations: water electrolysis (industrial electricity)		Hydrogen gas tank	Hydrogen gas	• Low energy efficiency • Has high concentration of air pollutant emission
10	Refueling stations: water electrolysis (Valley electricity)		Hydrogen gas tank	Hydrogen gas	• Low energy efficiency • Has high concentration of air pollutant emission • The highest cost
11	Central factory: methanol synthesis via natural gas	Methanol tank by truck	Methanol tank	Methanol Reforming on board	• 2nd lowest emission of air pollutant • The lowest costs in term of material

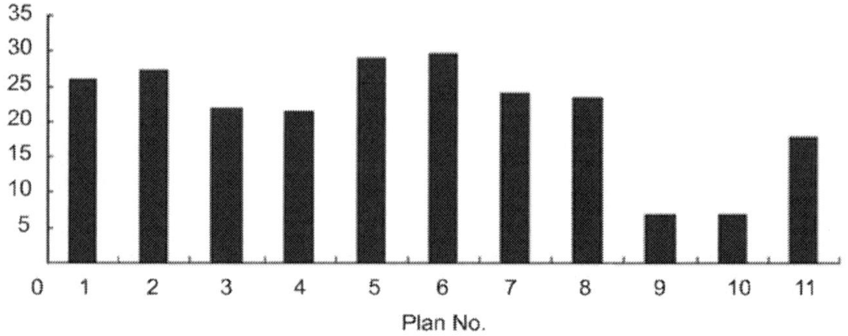

Figure 7: The total energy efficiency in percentage [63].

In terms of pollutant emission from the plant operation can be concluded as per Fig. 8. The ranking of four methods of producing hydrogen in environmental performance is NGSR, methanol reforming on-board, coal gasification, and water electrolysis. And as for the rank of four methods of storing and transporting hydrogen in environmental performance is, hydrogen gas by pipeline, hydrogen gas by cylinder, liquid hydrogen, and hydride [63].

In Fig. 9 the ranking of four methods of producing hydrogen in economic performance is: methanol reforming on-board, coal gasification, NGSR, and water electrolysis. The four methods of storing and transporting hydrogen in economic performance are ranked as hydrogen gas by cylinder, liquid hydrogen, and hydrogen gas by pipeline, and hydride [59]. Plan 10 is more advantageous than Plan 9 in economic performance because valley electricity is much cheaper than industrial electricity, although this will cause more equipment of water electrolysis to be needed in Plan 10 than that in Plan 9 owing to shorter work- hour in Plan 10 [64].

Overall the best plant in terms of energy performance is the plant that uses the combination of coal gasification and pipeline with total efficiency 30%. From the environmental aspect, perfor- mance is the combination of the usage NGSR and pipeline emitted the lowest emission of dangerous gas pollutant and regarding economic aspect, performance is the usage of methanol reforming on-board, gives the lowest cost [66].

Public Acceptability of Hydrogen Fuelling Station

In development of hydrogen, the public acceptability and behaviour towards hydrogen fuelling station should also be considered. In previous study, a survey was conducted regarding hydrogen fuel stations and safety at three locations, which are in Back Yard, Greater Stavanger and London [67]. From the survey, it shows that most of people in Greater Stavanger support the fuelling station, whereas for London, it shows that people need more information about hydrogen but still has high percentage of people who support hydrogen fuelling station as compared to people who are opposed and indifferent to hydrogen fuelling stations [17,68].

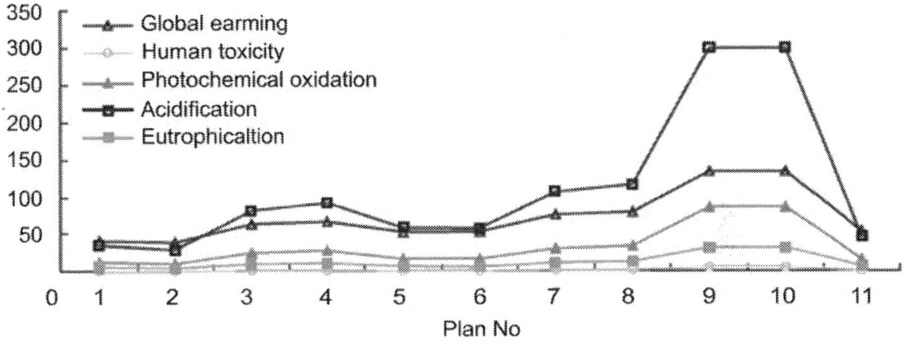

Figure 8: The standard indexes of classified environment effect [63].

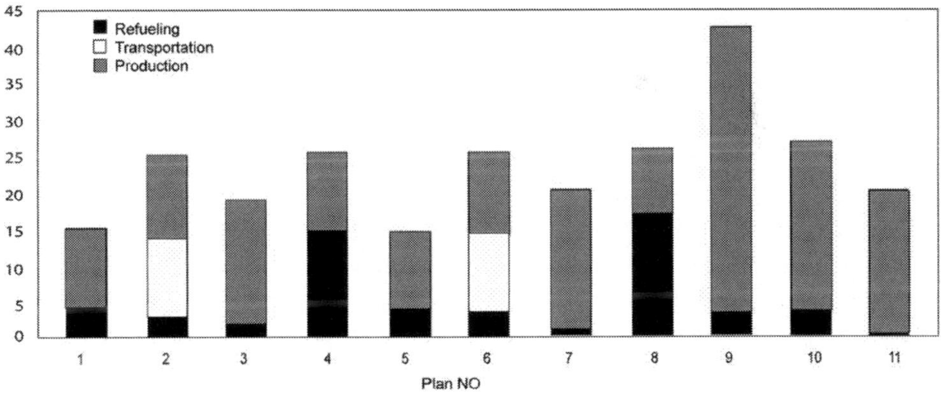

Figure 9: The constitution of hydrogen cost (Yuan RMB/kg H_2) [63].

The main point that helps people to accept hydrogen is the knowledge and awareness of hydrogen; that can be seen in Fig. 10. Most of the people in London need more information regarding hydrogen, whereas for Greater Stavanger and Back Yard, they already have the knowledge about hydrogen and give high percentage of support. As for London the percentage may be changed if they know more about hydrogen [69]. These conclu- sions are supported by another study done by a survey that had been conducted before and after that people experienced with the hydrogen vehicle. Another part of the study is done by looking at peoples' response towards how far they would travel for hydro- gen fuelling station? It also shows a positive feedback from people. In terms of fuelling station location people who live nearby the station give quite high support towards the hydrogen station implementations [70,71].

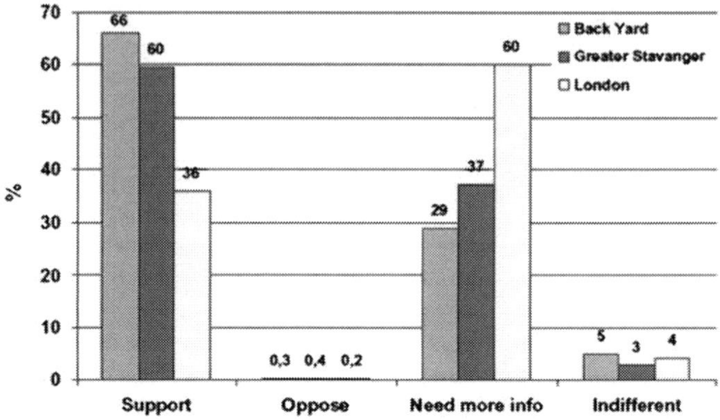

Figure 10: People respond towards hydrogen vehicle [68].

Overall, the media should play an important role in order to give the information regarding hydrogen in terms of environmental aspect, technological progress and regional success. This factor may contribute to the acceptability of hydrogen vehicles people [72]. In the development of hydrogen technology, safety measure- ment should also be considered, even though the society does not take it into their main consideration. Accidents may occur and will affect people judgments towards hydrogen vehicle [68,71].

Another study shows that in order to develop hydrogen technology, we should consider the usage of electricity needed for hydrogen production, to make sure that the hydrogen fuel can be competitive to other fuel options. Either than that, based on fuel price in certain country, hydrogen can compete with gasoline price under conditions of electricity price and fuel taxes. And lastly the storage of hydrogen is the main technical issue for hydrogen production, and the best solution maybe by developing inexpensive hydrogen storage tanks [71,73].

Life Cycle of Hydrogen

Life cycle is a process of a product or a service from its extraction of natural sources to its disposal. In previous study the life cycle of hydrogen is assessed through comparison, by comparing life cycle of gasoline from crude oil, hydrogen from natural gas and two types of renewable energies, which are solar and wind energies.

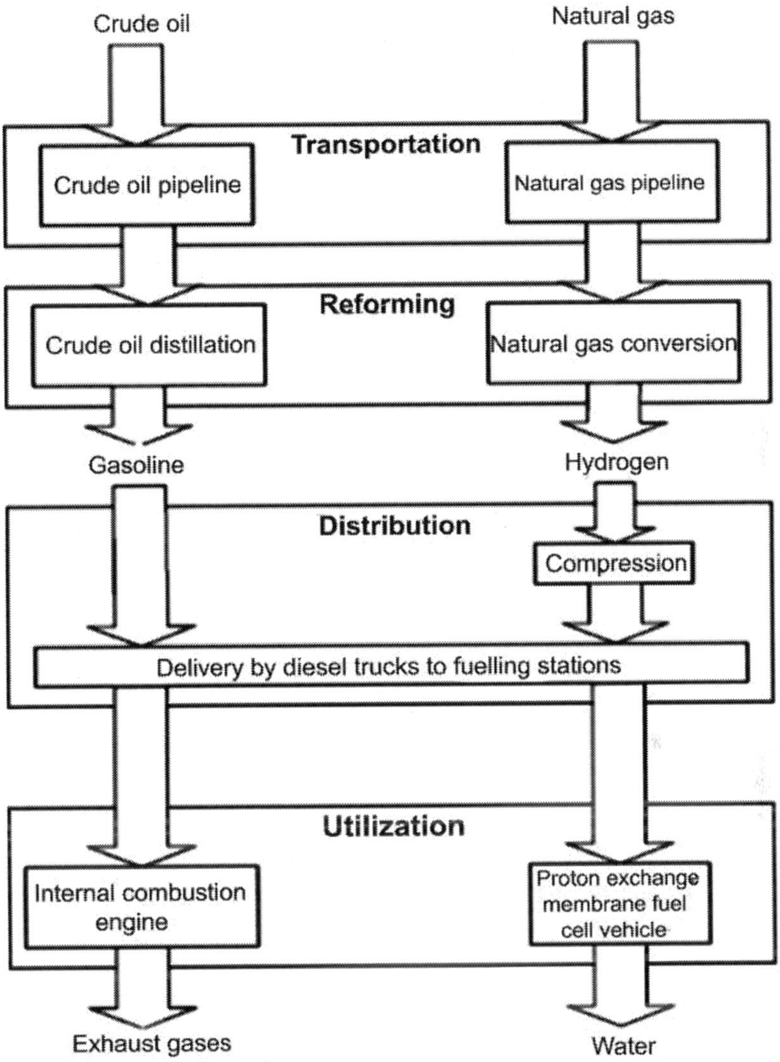

Figure 11: Production of gasoline and hydrogen from fossil fuel [74].

Fig. 11 shows the life cycle process for crude oil and natural gas whereas, Fig. 12 shows the renewable energy life cycle to produce hydrogen [74]. From Fig. 11, the extraction from natural sources is fossil fuels. These then are transport to reforming plant by using pipeline. Reforming is the production process of gasoline and hydrogen from fossil fuels. This process produces gasoline and hydrogen, both fuels are then transported or distributed to fuelling station by using tank trucks, but as for hydrogen it needs additional pressure to be stored in tank, hence adding some additional energy and also material for the tanks. The final stage is the usage by the consumers and the emission of both fuels [74,75]. As for renewable energy in Fig. 12, its natural sources are wind and solar. Both solar and wind energies generate electricity via photovoltaic element and wind turbine, respectively. The electricity is transported directly to fuelling station to produce hydrogen using chemical reactions (electrolysis). The hydrogen produced then is compressed before it can be stored. And the final stage is same as fossil fuels, the utilization or the usage of the consumables [74,76].

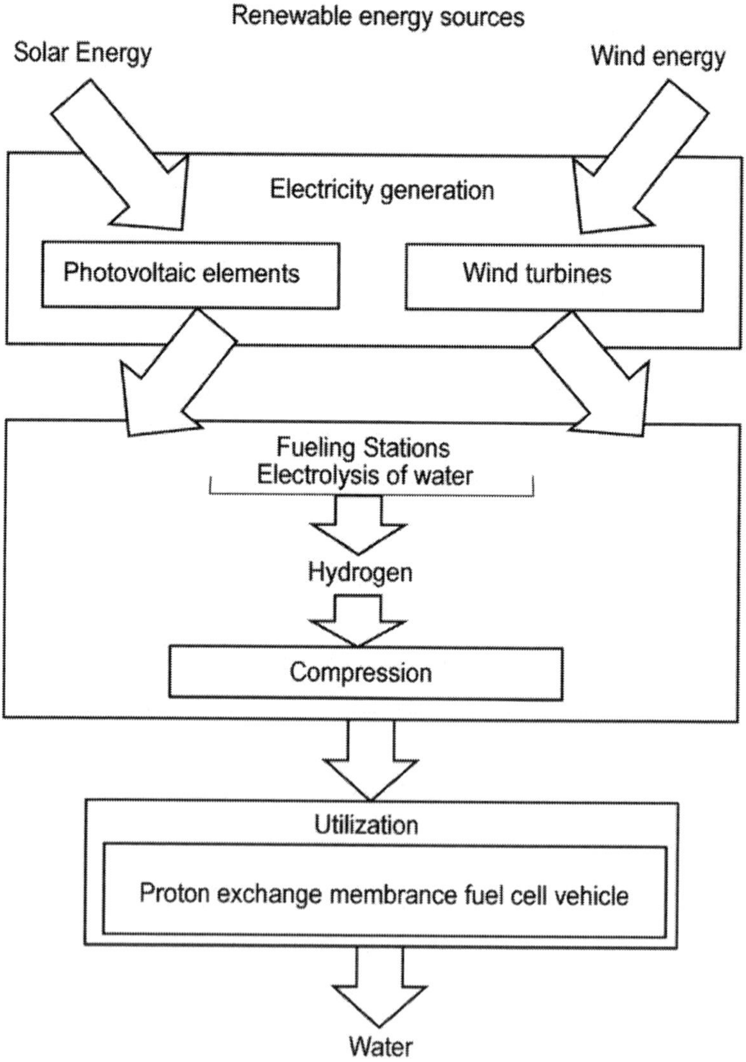

Figure 12: Production of hydrogen from renewable energy [74].

Fig. 13 shows that the energy consumption, to produce gaso- line, is less compared to hydrogen via natural gas team reforming. Whereas for renewable energy, wind energy uses low energy consumption to produce hydrogen and the solar is the less efficient due to the energy consumption needed to produce hydrogen fuels [62]. Fig. 14 shows that hydrogen and gasoline produce almost the same amount of Carbon Dioxide. The hydro- gen emits high volume of CO_2 during the production process. As for gasoline, the high emission comes from the fuel utilization. For renewable energies both show low emission of CO_2, but the lowest is from wind energy [74].

The gasoline production from crude oil has better efficiency compared to production of hydrogen using natural gas from all of above methods in producing fuels. The emission of both, hydro- gen fuel from natural gas and gasoline from crude oil show no significant difference, both has high emissions of greenhouse gases compared to renewable energy source [62]. The high emission comes from the production process. In terms of cost to produce hydrogen via natural gas is about five times less than the cost to produce hydrogen via wind energy, due to the construc- tion materials of the technology, but it can be reduced if further study is done in terms of reducing the material used in the construction of the wind turbine and also the improvement in terms of electricity generated by the turbine [74–76].

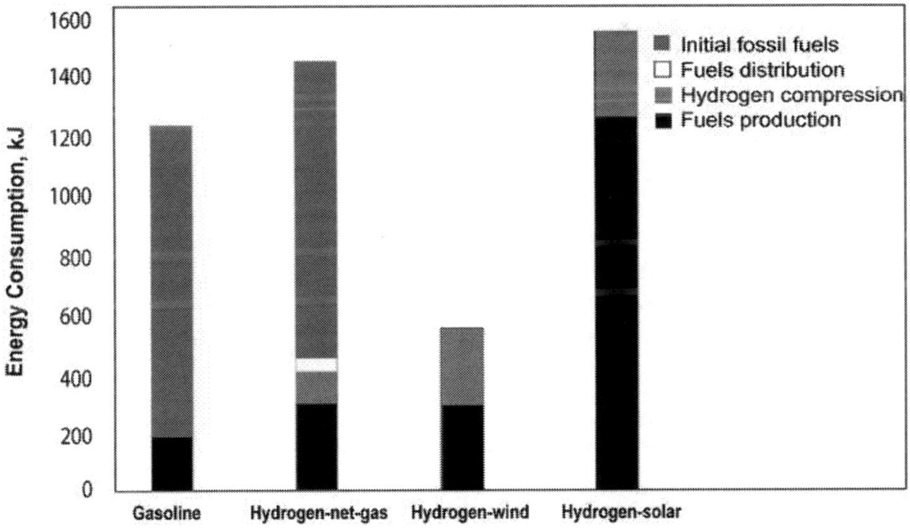

Figure 13: Energy consumption by type of process [74].

Overall, the advantages, disadvantages and improvements of hydrogen in internal combustion engine are tabulated in Table 12.

HYDROGEN PRODUCTION

There are many ways to generate hydrogen as an energy carrier and the sources are so abundant in this world. The biggest part of today's 500 billion cubic meters hydrogen sold worldwide is generated from fossil sources (natural gas, oil), or is obtained as by-product-hydrogen in chemical processes [77]. There are many processes of chemical processes for fuel cell vehicles such as small reformer, steam reforming and partial oxidation; gasification. Besides those on-board hydrogen productions, hydrogen can also be produced by electrolysis. There are many types of electrolysis such as alkaline water electrolysis, Proton-Exchange-Membrane (PEM), water electrolysis and High Temperature Electrolysis [78]. Other than water electrolysis, the hydrogen can also be produced by biomass gasification, which is also one of the renewable resources. Different types of hydrogen productions have their own source and it varies in terms of system applications as well. The best method to produce hydrogen is the one which has simplest process, easily to get the main sources, low cost and environmentally safe [6].

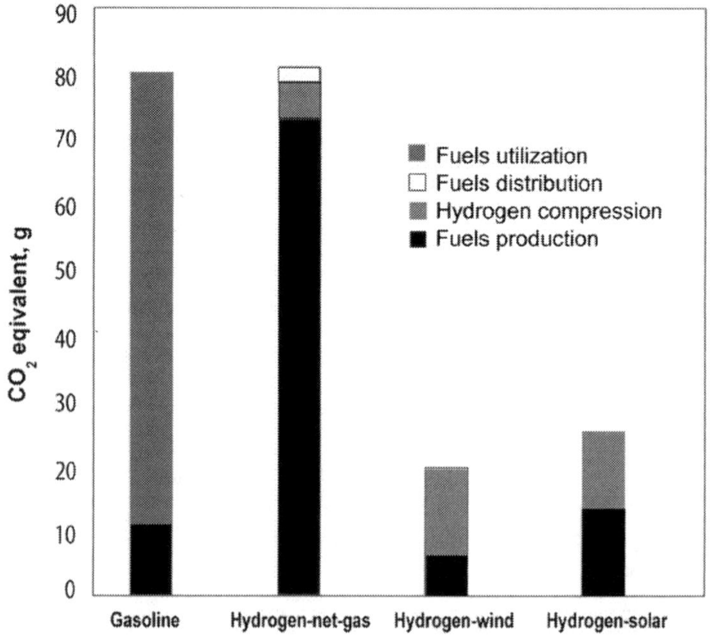

Figure 14: Greenhouse gases emitted by type of hydrogen production [74].

Natural Gas to Hydrogen

Natural gas is considered as a fossil fuel since it is formed from tiny sea animal and plants that died 200 to 400 million years ago. Raw natural gas consists of many different gases with the main gas is methane that is mixed with heavier hydrocarbon and carbon dioxide. Steam reforming is the process to convert natural gas to hydrogen. In high temperature steam, the hydrogen atoms separate from the carbons atoms in methane (CH_4). The reactions are reversible in nature, first is an endothermic reaction, which is the reaction that consumes heat to produce synthetic gases as H_2 and CO. During the reaction process, the methane reacts with steam at 750 1C to 800 1C with the pressure 3 bar to 25 bar [79]. The second reaction which, known as a water gas shift reaction, is exothermic that mildly produces heat. This process occurs in two stages, consisting of a high temperature shift (HTS) at 350 1C and a low temperature shift (LTS) at 190–210 1C [79]. The chemical equations to produce hydrogen by natural gas reforming are shown as following [80,81]:

$$CH_4 + H_2O\ (\text{steam}) \rightarrow 3H_2 + CO \quad \ldots(5)$$

Table 12: Positive features, limitations and improving the operational of hydrogen for engine application [29].

Positive features of hydrogen for engine application	Limitation associated with hydrogen engine applications	Improving the operational features of SI hydrogen engines
Less cyclic variations—This leads to a reduction in emissions, improved efficiency, and quieter and smoother operation.	Engines fuelled with H_2 suffer from reduced power output, due mainly to the very low heating value of H_2 on volume basis.	Employ lean mixtures with wide-open throttle. (To apply optimal variable partial throttling at extremely lean mixtures to effect better engine performance)
H_2 engines are more amenable to high-speed engine operation mainly due to the associated fast burning rates.	The mass of the intake air is reduced for any engine size because of the relatively high stoichiometric H_2 to air ratio.	Uniquely compatible and specially designed turbochargers need to be used for hydrogen engine applications.
Moderately high compression ratio operation is possible with lean mixtures of H_2 in air, which permits higher efficiencies and increased power output	There are serious potential operational problems associated with the uncontrolled pre-ignition and backfiring into the intake manifold of H_2 engines.	Higher compression ratios can be applied satisfactorily to increase the power output and efficiency, because of the relatively fast burning characteristics of the very lean H_2–air mixtures.
The reaction rates of H_2 are sensitive to the presence of a wide range of catalysts. This feature helps to improve its combustion and the treatment of its exhaust emissions.	The high burning rates of H_2 produce high pressures and temperatures during combustion in engines when operating near stoichiometric mixtures. This may lead to high NO_x emissions.	Carefully controlled cooling of EGR can be applied for knock avoidance and control. For lean mixture operation with H_2 suitably heated exhaust gas recirculation can be used.
The thermodynamic and heat transfer characteristics of H_2 tend to produce high compression temperatures that contribute to improvements in engine efficiency and lean mixture operation.	Hydrogen engine operation may be associated with increased noise and vibrations due mainly to the high rates of pressure rise resulting from fast burning.	Time injection of PFI or DI need to be optimized for injection duration, timing and pressure. This is important especially for the avoidance of pre-ignition and backfiring. Provision of some water injection when needed can be also made
H_2 high burning rates make the H_2 fuelled engine performance less sensitive to changes to the shape of the combustion chamber, level of turbulence and the intake charge swirling effect.	Great care is needed to avoid materials compatibility problems with hydrogen applications in engines.	Optimum spark ignition characteristics in terms energy, spark plug gap size and material, plug geometry, electrical insulation etc. need to be employed.
The gas is highly diffusive and buoyant which make fuel leaks disperse quickly, reducing the fire and explosion hazards associated with H_2 engine operation.	Hydrogen requires a very low ignition energy, which leads to uncontrolled pre-ignition problems.	Further improvement in performance can be obtained by having the design features of the combustion chamber and its surfaces suitably optimized for H_2 operation.
	There is an increased potential for undesirable corrosion and lubricating oil contamination due to exhaust water vapour condensation.	Variable valve timing needs to be incorporated and optimized to effect higher volumetric efficiency and better control of EGR.

$$CO + H_2O\ (steam) \rightarrow CO_2 + H_2 \quad \ldots(6)$$

Nowadays, the majority of hydrogen produced worldwide is accomplished by steam methane reforming. The efficiency of steam reforming process is about 65–75% among the highest of current commercially available products [79]. Natural gas is the best method to produce hydrogen as it is convenient, easy to handle, and high hydrogen-to-carbon ratio. Fig. 15 shows one of the hydrogen applications as reforming the natural gas. The hydrogen can be used in fuel cell in order to generate electricity.

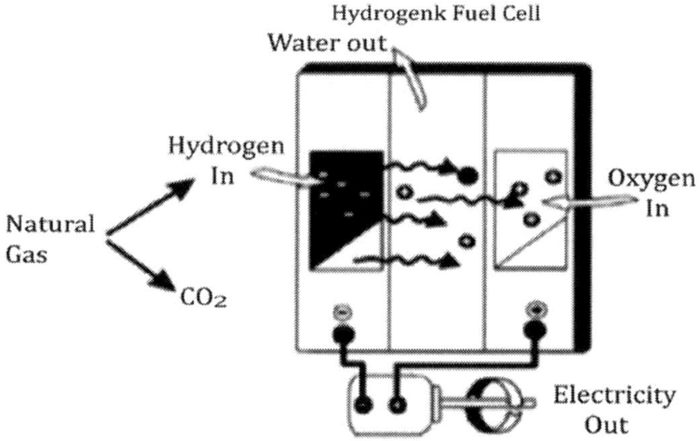

Figure 15: Hydrogen from natural gas for fuel cell application [81].

Coal Gasification

Through coal gasification with the addition of carbon capture technology, high volume stream of hydrogen can be produced. In gasification process, the mixing of pulverized coal with an oxidant, heated to about 1800 1C have a very hot synthesis (syngas). The syngas consists of hydrogen, carbon monoxide, carbon dioxide, other gases and particle. To remove the other gases and particle, the syngas is cooled and cleaning process is proceeded.

During the cleaning of syngas, any particulate such as mercury, sulphur, trace contaminants and foreign matter are removed. Then, the syngas reacts with steam by a process called water gas shift reaction to produce more hydrogen and carbon dioxide as final product as in Fig. 16. The hydrogen then can be used as a vehicle fuel or to generate electricity for other purposes such as power plant and industry. Meanwhile, as carbon dioxide has also been produced at the end, a technology is needed to decrease or at least sustain the amount of carbon dioxide emissions in the atmosphere. Nowadays, carbon dioxide emissions can be reduced near to zero when applying carbon capture storage and seques- trations technologies [82,83]. By increasing the efficiency of operation in the gasification process, the pollutants can also be reduced significantly.

Carbon capture and storage concept is used for capturing and storing the carbon dioxide permanently that has been produced from coal gasification or any other industrial activity [84]. The carbon is stored naturally in the earth's terrestrial biosphere (in forests, soil, and plants) and ocean reservoirs (via the ocean carbon cycle), from which it is cyclically released and absorbed [82]. The percentage of carbon dioxide emissions to atmosphere can be reduced is about 90% by implementing the concept of carbon capture and storage in pulverized coal plant.

Electrolysis

Fossil fuel energy sources such as natural gas, coal and petroleum are not sustainable as these are depleting and there are severe damages to the environment because of the activities to produce hydrogen. In order to make a global sustainability and stability, renewable energy such as electrolysis and biomass gasification needs to be commercialized, and replaces the use of fossil fuel significantly [85].

To operate the electrolysis process, the electricity needed, can be generated either from fossil fuels or renewable energy such as solar power. In water electrolysis process, the hydrogen produced is clean, with high purity and is as simple as using electricity, generated by fossil fuels. Besides using electricity as the source, the hydrogen can also be produced by photo catalytic water splitting. This technology is still in experimental stage due to low efficiency and high cost [86]. At present, only about 5% of hydrogen in the world is produced by water electrolysis [87].

Currently, the best solution to the high cost of electrolysis process is by using sustainable sources such as solar, wind and or nuclear. In this paper, renewable energy as the source for electrolysis will be discussed. There are many types of electrolysis such as alkaline water electrolysis, polymer electrolyte membrane (PEM) electrolysis, solid oxide electrolysis, and photo-electrolysis [87,88]. The difference among the electrolysis systems is the source of power to conduct the electrolysis, the constructions, conversion efficiency, and availability in industries.

Alkaline water electrolysis is one of the easiest methods for hydrogen production because of its simple construction. A basic water electrolysis unit consists of an anode, a cathode, power supply and an electrolyte [88]. The molecules of water (H_2O) can be split to form pure hydrogen and oxygen by using electricity. However, when the electricity used, is generated by burning fossil fuels, the pollutant emissions cannot be avoided. But producing hydrogen by using renewable energy source, the efficiency is around 68% [89]. For application in transportation sector, the electrolyzers, used in the electrolysis process, can be reduced in size to suit the fuel cell vehicles that give an important advantage in the development of FCV market.

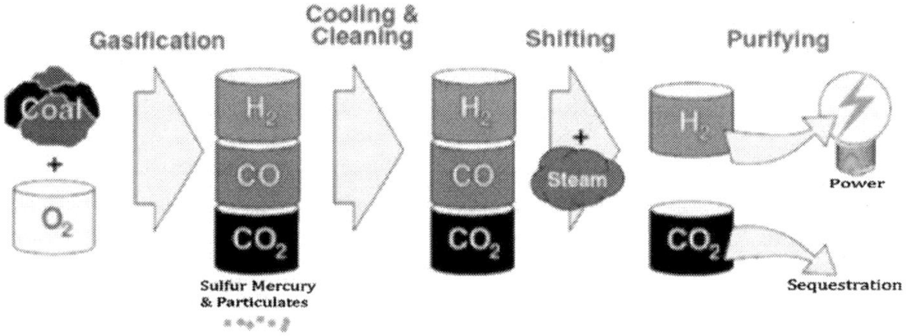

Figure 16: Coal gasification process [82].

Table 13 shows the different electrolysers that can be used to produce hydrogen gas. The different electrolytes are used in electrolysis and the efficiency in producing the hydrogen. Besides that, the maturity or the level of acceptance in the current market also varies. The PEM electrolyser has the highest efficiency of 65–85% and can be one of the replacement to the existing alkaline electrolyser. Photoelectrolysis technology which uses solar power as its main source needs more research and devel- opment in future to have high efficiency [90].

Table 13: Efficiency comparison among different types of electrolyser [88].

Technology	Efficiency (%)	Maturity
Alkaline electrolyser	59–70	Commercial
PEM electrolyser	65–82	Near term
Solid oxide electrolysis cells	40–60	mediate term
Photo electrolysis	2–12	long term

Biomass Gasification

Biomass can be a fuel source, derived from plant and animal wastes. From biomass, the natural gas (methane) can be obtained. Actually, methane can be obtained naturally as the waste; organic matter decays. Landfills are the places, where methane can be collected. The methane gas is used for heating and producing electricity.

Biomass gasification is one of the most mature technologies to produce syn-gas. This technology is however very expensive due to high energy requirements and inherent energy losses in biomass gasification. Biomass gasification means incomplete combustion of biomass that produces combustible gases consist- ing of carbon monoxide (CO), hydrogen (H_2) and methane (CH_4). The mixture of combustible gases is also known as producer gas. Producer gas can be used to run both combustion engines either compression or spark ignition. The production of producer gas is called gasification. Gasification is partial combustion of biomass and is reacted in gasifier at 1000 1C [91].

The gasification process occurs in a gasifier involving four processes; (a) drying the fuel (b) pyrolysis (c) combustion (d) reduction [91]. Sometimes, the processes are overlapping but the processes can still be classified, happening at different zones and temperatures, based on the different chemical and thermal reactions.

Photolytic Processes

Photolytic processes use light energy to split water into hydrogen and oxygen. Currently in the very early stages of research, these processes offer long-term potential for sustainable hydrogen production with low environmental impact. Two pro- cesses to produce hydrogen are used in photolytic processes, namely, photobiological water splitting and photoelectrochemical water splitting.

Photobiological Water Splitting

In this process, sunlight and specialized microorganisms are sources for hydrogen production, such as green and cyanobac- teria. Hydrogen is produced as by-product by these microorgan- isms as a by-product of their natural metabolic process, just as plants produce oxygen during photosynthesis. Photobiological water splitting is a long-term technology. At present, for efficient and commercial hydrogen production, the microbes split water very slow, which is to be used.

There are many ways under research by scientists to modify the microorganisms and to identify other naturally occurring microbes, which can produce hydrogen at higher rates. Photobiological water splitting offers long-term potential for sustainable hydrogen production with low environmental impacts, even though it is in the very early stages of research [92].

Photoelectrochemical Water Splitting

Sunlight and specialized semiconductors called photoelectrochemical materials are used to produce hydrogen, in this process. In the photoelectrochemical (PEC) system, light is directly used to dissociate water molecules into hydrogen and oxygen by the semiconductor. Different semiconductor materials work at parti- cular wavelengths of light and energies.

Research focuses on finding semiconductors with the correct energies to split water that are also stable when in contact with water. Photoelectrochemical water splitting offers long-term potential for sustainable hydrogen production with low environ- mental impacts, even though it is in very early stages of research [93].

CONCLUSION

The crucial outcomes of this study are summarized below:

- Hydrogen in internal combustion engines has many advantages in terms of combustive properties but it needs detailed consideration of engine design to avoid abnormal combustion, which is the major problem in hydrogen engine. This, as a result can improve engine efficiency, power output and reduce NO_x emissions.
- In fuel cell vehicles, the hydrogen purity can affect the performance of the fuel cell vehicles. This impurity comes from the poisoning of the sulphur during production process. From the environmental aspects, the emission of fuel cell is low as compared to conventional vehicles but as penalty, fuel cell vehicles need additional space and weight to install the battery and storage tank, thus increases it production cost.
- The cost and also the efficiency of the hydrogen plant depend on the electricity tariff and the sources for producing hydro- gen. If the location is near with its natural resources, it will help in reducing its cost, so for the development of hydrogen plant, location with its sources should be considered.

- The acceptability of hydrogen technology by people is through the knowledge and also the awareness of the hydrogen benefits towards environment and human life. Recent study shows that people still do not have the information of hydro- gen. Media role in introducing hydrogen technology to citizens is crucial in order to get support of people in development of hydrogen technology.
- There are many ways to generate hydrogen as an energy carrier and the sources are in abundance. Mainly it is produced from fossil fuels and as by-product hydrogen in chemical processes. Different types of hydrogen productions have their own source and it varies in terms of system applications as well.
- The best method to produce hydrogen is the one which has simplest process, easily to get the main sources, low cost and environmentally safe.
- The study in the production methods, vehicle performance, plant performance, infrastructure availability, emissions and air pollution is needed, before the hydrogen fuel and vehicles can be commercialized and compete with other type of fuels.

cknowledement

The authors would like to acknowledge the Ministry of Higher Education of Malaysia and The University of Malaya, Kuala Lumpur, Malaysia for the financial support under UM.C/HIR/ MOHE/ENG/15 (D000015-16001).

REFERENCES

Abbasi T, Abbasi SA. 'Renewable' hydrogen: prospects and challenges. Renewable and Sustainable Energy Reviews 2011;15(6):3034–40.

Cotrell J, Pratt W. Modeling the Feasibility of Using Fuel Cells and Hydrogen Internal Combustion Engines in Remote Renewable Energy Systems 2003.

Liu PF, et al. Numerical simulation and optimal design for composite high-pressure hydrogen storage vessel: a review. Renewable and Sustainable Energy Reviews 2012;16(4):1817–27.

Balat M. Hydrogen in fueled systems and the significance energy sourcesof hydrogen in vehicular transportation. Energy Sources, part B 2007;2:49–61.

Farrauto RJ. 3 From the internal combustion engine to the fuel cell: moving towards the hydrogen economy. In: Masakazu Anpo MO, Hiromi Y, editors. Studies in surface science and catalysis. Elsevier; 2003. p. 21–9.

Silveira JL, et al. The benefits of ethanol use for hydrogen production in urban transportation. Renewable and Sustainable Energy Reviews 2009;13(9): 2525–34.

Ganesan V. Internal combustion engines. New Delhi: Tata McGraw-Hill Publishing Company Limited; 2003.

Saanum, I Experimental Studies of Hydrogen as a Fuel Additive in Internal Combustion Engines, in Department of Energy and Process Engineering. 2008, Norwegian University of Science and Technology.

Sobrino FH, Monroy CR, Pe´ rez JLH. Critical analysis on hydrogen as an alternative to fossil fuels and biofuels for vehicles in Europe. Renewable and Sustainable Energy Reviews 2010;14(2):772–80.

Guo LS, Lu HB, Li JD. A hydrogen injection system with solenoid valves for a four-cylinder hydrogen-fuelled engine. International Journal of Hydrogen Energy 1999;24(4):377–82.

Soberanis MAE, Fernandeza AM. A review on the technical adaptations for internal combustion engines to operate with gas/hydrogen mixtures. Inter- national Journal of Hydrogen Energy 2010;35(21):12134–40.

Verhelst S, Thomas W. Hydrogen-fueled internal combustion engines. Progress in Energy and Combustion Science 2009;35(6):490–527.

Momirlan M, Veziroglu TN. Current status of hydrogen energy. Renewable and Sustainable Energy Reviews 2002;6(1–2):141–79.

Bauer CG, Forest TW. Effect of hydrogen addition on the performance of methane-fueled vehicles. Part I: Effect on S.I. engine performance. Interna- tional Journal of Hydrogen Energy 2001;26(1):55–77.

Wierzba I, Kilchyk V. Flammability limits of hydrogen–carbon monoxide mixtures at moderately elevated temperatures. International Journal of Hydrogen Energy 2001;26(6):639–43.

Roy MM, et al. An experimental investigation on engine performance and emissions of a supercharged H_2-diesel dual-fuel engine. International Journal of Hydrogen Energy 2010;35(2):844–53.

Midilli A, et al. On hydrogen and hydrogen energy strategies II: Future projections affecting global stability and unrest. Renewable and Sustainable Energy Reviews 2005;9(3):273–87.

White CM, Steeper RR, Lutz AE. The hydrogen-fueled internal combustion engine: a technical review. International Journal of Hydrogen Energy 2006;31:1292–305.

Willard, WP, Engineering Fundamentals of the Internal Combustion Engine.

Al-Baghdadi MARS. Effect of Compression Ratio, Equivalence Ratio and Engine Speed on the Performance and Emission Characteristics of a Spark Ignition Engine Using Hydrogen as a Fuel. 2004 2004;29:2245–60.

Balat M. Potential importance of hydrogen as a future solution to environmental and transportation problems. International Journal of Hydrogen Energy 2008;33:15.

Liu X-h, et al. Backfire prediction in a manifold injection hydrogen internal combustion engine. International journal of hydrogen energy 2008;33(33): 3847–55.

Al-Baghdadi, MARS Measurement and prediction study of the effect of ethanol blending on the performance and pollutants emission of a four- stroke spark ignition engine Proceedings of the institution of mechanical engineers, Part D: Journal of automobile engineering; 2008. 222: p. 859–873.

Heywood, J Internal Combustion Engine Fundamentals. 1989, New York.

Li H, Karim GA. Knock in spark ignition hydrogen engines. International Journal of Hydrogen Energy 2004;29(8):859–65.

Sadiq Al-Baghdadi Maher AR. Effect of compression ratio, equivalence ratio and engine speed on the performance and emission characteristics of a spark ignition engine using hydrogen as a fuel. Renewable Energy 2004;29(15): 2245–60.

Saxena RC, et al. Thermo-chemical routes for hydrogen rich gas from biomass: a review. Renewable and Sustainable Energy Reviews 2008;12(7): 1909–27.

Kirchweger W, et al. Applications of the LIF method for the diagnostics of the combustion process of gas-IC-engines. Experiments in Fluids 2007;43: 329–40.

Lee SJ, Yi HS, Kim ES. Combustion characteristics of intake port injection type hydrogen fueled engine. Hydrogen Energy 1995;20(4):317–22.

Szwaja S, Bhandary KR, Naber JD. Comparisons of hydrogen and gasoline combustion knock in a spark ignition engine. International Journal of Hydrogen Energy 2007;32(18):5076–87.

Hwang JJ. Review on development and demonstration of hydrogen fuel cell scooters. Renewable and Sustainable Energy Reviews 2012;16(6):3803–15.

Yousufuddin S, Mehdi SN, Masood M. Performance evaluation of a hydrogen–ethanol fuelled engine. International Journal of Energy Technology and Policy (IJETP) 2009;7:2.

Ghazi AK. Hydrogen as a spark ignition engine fuel. International Journal of Hydrogen Energy 2003;28(5):569–77.

Kleijn R, van der Voet E. Resource constraints in a hydrogen economy based on renewable energy sources: an exploration. Renewable and Sustainable Energy Reviews 2010;14(9):2784–95.

Verhelst, S, Verstraeten S, and Sierens R. A comprehensive overview of hydrogen engine design features. Proceedings of the institution of mechan- ical engineers, Part D: Journal of automobile engineering; 2007. 221(8): p. 911–920.

Das L. Near-term introduction of hydrogen engine for automotive and agricultural application. Institute Journal of Hydrogen Energy 2002;27(5): 479–87.

Mohammadi A, et al. Performance and combustion characteristics of a direct injection SI hydrogen engine. International Journal of Hydrogen Energy 2007;32(2):296–304.

Adnan R, Masjuki HH, Mahlia TMI. Performance and emission analysis of hydrogen fueled compression ignition engine with variable water injection timing. Energy 2012(0).

Liu J, et al. Numerical study of hydrogen addition to DME/CH4 dual fuel RCCI engine. International Journal of Hydrogen Energy 2012;37(10):8688–97.

Verhelst S, Sierens R. Hydrogen engine-specific properties. International Journal of Hydrogen Energy 2001;26(9):987–90.

Sun Z-y, et al. Research and development of hydrogen fuelled engines in China. International Journal of Hydrogen Energy 2012;37(1):664–81.

SinghYadav V, Soni SL, Sharma D. Performance and emission studies of direct injection C.I. engine in duel fuel mode (hydrogen–diesel) with EGR. Interna- tional Journal of Hydrogen Energy 2012;37(4):3807–17.

Mariani A, Morrone B, Unich A. Numerical evaluation of internal combustion spark ignition engines performance fuelled with hydrogen—natural gas blends. International Journal of Hydrogen Energy 2012;37(3):2644–54.

Wang S, et al. Performance of a hydroxygen-blended gasoline engine at different hydrogen volume fractions in the hydroxygen. International Journal of Hydrogen Energy 2012(0).

Echekki T, Gupta KG. Hydrogen autoignition in a turbulent jet with preheated co-flow air. International Journal of Hydrogen Energy 2009;34(19):8352–77.

Tiwari GP, et al. A study of internal hydrogen embrittlement of steels. Materials Science and Engineering 2000;286(2):269–81.

Moreno F, et al. Efficiency and emissions in a vehicle spark ignition engine fueled with hydrogen and methane blends. International Journal of Hydrogen Energy 2012(0).

Nieminen J, D'Souza N, Dincer I. Comparative combustion characteristics of gasoline andhydrogen fuelled ICEs. International journal of hydrogen energy 2010;35(10):5114–23.

Aleiferis PG, Rosati MF. Flame chemiluminescence and OH LIF imaging in a hydrogen-fuelled spark-ignition engine. International Journal of Hydrogen Energy 2012;37(2):1797–812.

Balat M, Balat M. Political, economic and environmental impacts of biomass-based hydrogen. International Journal of Hydrogen Energy 2009;34(9): 3589–603.

Bose PK, Maji D. An experimental investigation on engine performance and emissions of a single cylinder diesel engine using hydrogen as inducted fuel and diesel as injected fuel with exhaust gas recirculation. International Journal of Hydrogen Energy 2009;34(11):4847–54.

Shudo T, Omori K, Hiyama O. NO_x reduction and NO_2 emission characteristics in rich-lean combustion of hydrogen. International journal of hydrogen energy 2008;33(17):4689–93.

James WH. NO_x emission and performance data for a hydrogen fueled internal combustion engine at 1500 rpm using exhaust gas recirculation. International Journal of Hydrogen Energy 2003;28(8):901–8.

HICE. Hydrogen Internal Combustion Engine. 2010; Available from: /http://www.ika.rwth-aachen.de/r2h/index.php/ICES.

Energy, U.S.D.o. Hydrogen use in internal combustion engines 2010; Available from: /http://www1.eere.energy.govS.

Granovskii M, Dincer I, Rosen MA. Economic and environmental comparison of conventional, hybrid, electric and hydrogen fuel cell vehicles. Journal of Power Sources 2006;159(2):1186–93.

Sadiq Al-Baghdadi Maher AR Development of a pre-ignition submodel for hydrogen engines Proceedings of the institution of mechanical engineers, Part D: Journal of automobile engineering; 2005. 219: p. 1203–1212.

Haseli Y, Naterer GF, Dincer I. Comparative assessment of greenhouse gas mitigation of hydrogen passenger trains. I nternati onal Journal of hydrogen energy 2008;33:1788–96.

Kothari R, Buddhi D, Sawhney RL. Comparison of environmental and economic aspects of various hydrogen production methods. Renewable and Sustainable Energy Reviews 2008;12(2):553–63.

Offer GJ, et al. Comparative analysis of battery electric, hydrogen fuel cell & hybrid vehicles in a future sustainable road transport system. Energy Policy 2010;38(1):24–9.

Zhou L. Progress and problems in hydrogen storage methods. Renewable and Sustainable Energy Reviews 2005;9(4):395–408.

Tanksale A, Beltramini JN, Lu GM. A review of catalytic hydrogen production processes from biomass. Renewable and Sustainable Energy Reviews 2010;14(1):166–82.

Feng W, et al. The future of hydrogen infrastructure for fuel cell vehicles in China and case of application in Beijing. International Joural of Hydrogen Energy 2004;29(4):355–67.

Zoulias EI, et al. Integration of hydrogen energy technologies in stand-alone power systems analysis of the current potential for applications. Renewable and Sustainable Energy Reviews 2006;10(5):432–62.

Hart, D, A Bauen, and J Cross, Hydrogen Energy & Fuel Cells. 2003, London, Luxembourg.

Ally J, Pryor T. Accelerating hydrogen implementation by mass production of a hydrogen bus chassis. Renewable and Sustainable Energy Reviews 2009;13(3):616–24.

Schmidtchen U, et al. Hydrogen aircraft and airport safety. Renewable and Sustainable Energy Reviews 1997;1(4):239–69.

Thesena G, Langhelle O. Awareness, acceptability and attitudes towards hydrogen vehicles and filling stations: a greater stravanger case study and comparisons with London. International Journal of Hydrogen Energy 2008;33(21):5859–67.

Midilli A, et al. On hydrogen and hydrogen energy strategies: I: Current status and needs. Renewable and Sustainable Energy Reviews 2005;9(3): 255–71.

Safari H, Jazayeri SA, Ebrahimi R. Potentials of NO_x emission reduction methods in SI hydrogen engine: simulation study. International journal of hydrogen energy 2009;34(2):1015–25.

Elliot M, et al. Behavirol response to hydrogen fuel cell vehicles and refueling: results of California drive clinics. International Journal of Hydrogen Energy 2009;34(20):8670–80.

Mirza UK, et al. A vision for hydrogen economy in Pakistan. Renewable and Sustainable Energy Reviews 2009;13(5):1111–5.

Prince-Richard S, Whalel M, Djilali N. A techno-economic analysis of decentralized electrolytic hydrogen production for fuel cell vehicles. Inter- national Journal of Hydrogen Energy 2005;30(11):1159–79.

Mikhail Granovskii ID, Rosen. Marc A. Life cycle assessment of hydrogen fuel cell & gasoline vehicles. International Journal of Hydrogen Energy 2006;159(2):1186–93.

Pamela LS, Margaret KM. Life cycle assessment of hydrogen production via natural gas reforming. National Renewable Energy Laboratory; 2001.

Pamela LS, Margaret KM. Life cycle assessment of renewable hydrogen production via wind/electrolysis. National Renewable Energy Laboratory; 2004.

Wikipedia, Hydrogen production, Retrieved on november 2010, Available at:/ http://en.wikipedia.org/wiki/Hydrogen_productions.

de Souza ACC, Silveira JL. Hydrogen production utilizing glycerol from renewable feedstocks—the case of Brazil. Renewable and Sustainable Energy Reviews 2011;15(4):1835–50.

Authority, N.Y.S.E.R.a.D., Hydrogen Production—Coal, New York State Energy Research and Development Authority. Retrieved on 09-09-2010, Available at:

/www.nyserda.orgS. Online 3/8/2010. 2010.

Besancon M, B, et al. Hydrogen quality from decarbonized fossil fuels to fuel cells. International Journal of Hydrogen Energy 2009;34(5):2350–60.

Anonymous. Secondary Energy Infobook 2008.

HPC, *Hydrogen Production from Coal.* The National Hydrogen Association, Retrieved on 05-09-2010 [Online] March 2010. /www. HydrogenAssociation. orgs., 2010.

Colella WG, Jacobson MZ, Golden DM. Switching to a U.S. hydrogen fuel cell vehicle fleet: the resultant changes in emissions, energy use, and greenhouse gases. Journal of Power Sources 2005;150:150–81.

Mazloomi K, Gomes C. Hydrogen as an energy carrier: prospects and

challenges. Renewable and Sustainable Energy Reviews 2012;16(5):3024–33.

Association, NH, Hydrogen Production from Coal. The National Hydrogen Association. /www.HydrogenAssociation.orgS. 13/9/2010. 2010.

Osterloh FE. Inorganic materials as catalysts for photochemical splitting of water. ChemInform 2008:39.

Wang M, Wang Z, Guo Z. Water electrolysis enhanced by super gravity field for hydrogen production. International Journal of Hydrogen Energy 2010;35(8):3198–205.

Zeng K, Zhang D. Recent progress in alkaline water electrolysis for hydrogen production and applications. Progress in Energy and Combustion Science 2010;36(3):307–26.

Ananthachar V, Duffy JJ. Efficiencies of hydrogen storage systems onboard fuel cell vehicles. Solar Energy 2005;78(5):687–94.

Wang HZ, et al. A review on hydrogen production using aluminum and aluminum alloys. Renewable and Sustainable Energy Reviews 2009;13(4): 845–53.

Goswami, Yogi. Alternative energy in agriculture. Gasbook Biomass Gasification 1986;2:83–102.

Oh S, et al. Photoelectrochemical hydrogen production with concentrated natural seawater produced by membrane process. Solar Energy 2011;85(9): 2256–63.

U.S. Department of Energy, A Prospectus for Biological H2 Production; Retrieved on 10-05-2012; Availble at: /http://www1.eere.energy.gov/hydro genandfuelcells/production/pdfs/photobiological.pdfs.

3

Hydrogen Fuel Cell Vehicle Study

Craig Davis

Ford Motor Co. (retired) and Michigan State University. Email: ldavis7@peo plepc.com

Dr. Seema Tiwari

Associate Professor Applied Science & General Engineering Army Institute of Technology, Pune.

Ankur Jain

Quantum University, Dehradun Highway, Mandawar, Roorkee, Uttarakhand 247167, India

Mr. John Abbas

IBIS Group, Lucknow, UP, India.

Prof. (Dr.) B. K. Sarkar (Patent Guru/ Expert)

MES, Geh Research, India, USA, Singapore, Japan, Austraila, Germany.

Prof. (Dr.) Reena Singh,

MES's, Pillai HOC College of Engineering &Technology, Rasayani Taluka Panvel, Dist, Navi Mumbai, Maharashtra 410207, India

INTRODUCTION

At the beginning of 2002, the Bush Administration announced the "FreedomCAR" initiative, an industry-government cooperative effort, to develop fuel cell vehicles. This prompted a subcommittee of the POPA Energy and Environment Committee to commence work on a report about fuel cells and FreedomCAR. The rationale for preparing such a report is that the topic is an important aspect of the nation's energy policy a topic that physicists justifiably feel competent to discuss. Previous POPA studies have been on nuclear energy, energy supplies, etc. Fuel cells are of interest to the physics community (e. g., see the recent Physics Today article by Joan Ogden [1]) and physicists are actively involved in research areas for potential hydrogen storage, such as carbon nanotubes. The materials aspects of fuel cells are especially within the purview of physicists. Overall systems considerations, wells-to-wheels energy efficiency, and related issues can benefit from analysis by physicists. In view of the high expectations for fuel-cell vehicles generated by the FreedomCAR initiative, it seems reasonable to examine what is reality and what is unsupported optimism. Of those who have read the Ogden article or popular-press fuel cell articles, some will want to know more. This report is a start on a balanced discussion that intends to educate, rather than persuade or advocate. The intended audience is POPA and the APS membership. [See Appendix A for fuel cell principles of operation and a schematic.]

The motivation for the FreedomCar initiative is to reduce U.S. dependence on imported petroleum, to reduce emissions of atmospheric pollutants, and to reduce CO_2emissions by improving fuel economy and/or by going to a hydrogen-based system. Since the transportation sector itself uses more oil than produced domestically (Fig. 1), FreedomCAR also addresses a serious national security issue.

The big three automotive manufacturers have publicly committed their companies to participation in the initiative. General Motors Chairman Jack Smith: "With the FreedomCAR program, we are taking a major step towards creating a future where the vehicle is no longer part of the energy and environmental debate." DaimlerChrysler CEO Dieter Zetsche: "FreedomCAR focuses on jointly developing technologies that are important to the entire automotive industry. This program allows us to continue to work together as an industry in a way that can make a difference." Ford Chairman and Chief Executive Officer William Clay Ford Jr.: "Our companies have made significant progress in reducing the environmental impact of our products. Our participation in FreedomCAR signifies our commitment to continue that progress."

FreedomCAR has the following technology-specific goals for 2010 [2] [See Appendix B for a complete set of goals and notes.]:

- To ensure reliable systems with costs comparable with conventional internal combustion engine/automatic transmission systems, future fuel cell powertrains should have
    - Electric propulsion system with a 15-year life capable of delivering at least 55 kW for 18 seconds and 30 kW in a continuous mode, at a system cost of $12/kW peak. [Note this pertains to electrical systems other than the fuel cell such as electric motors, controllers,etc.]
- A durable fuel cell power system (including hydrogen storage) that achieves 60% energy efficiency when operating at peak

power and that offers a 325 W/kg power density and 220 W/L operating on hydrogen. Cost targets are $45/kW by2010, $30/kW by 2015.

- To enable clean, energy-efficient vehicles operating on clean, hydrocarbon-based fuels powered by fuel cells, the goalis
  - Fuel cell systems, including a fuel reformer, that have a peak energy efficiency of 45% and meet or exceed emissions standards with acost target of $45/kW by 2010 and $30/kW in2015.
- To enable the transition to a hydrogen economy, ensure widespread availability of hydrogen fuels, and retain the functional characteristics of current vehicles, the goalsare
  - Demonstrated hydrogen refueling with developed commercial codes and standards and diverse renewable[1]and non-renewable energy sources. Targets: 70% energy efficiency well-to-pump; cost of energy from hydrogen equivalent to gasoline at market price, assumed to be $1.25 per gallon (2001dollars).
  - Hydrogen storage systems demonstrating an available capacity of 6 wt% hydrogen, specific energy of 2000 W-h/kg [pertains to storage system mass], and energy density of 1100 W-h/L at a cost of$5/kWh.
- To improve the manufacturing base, the goalis
  - Material and manufacturing technologies for high-volume production vehicles that enable and support the simultaneous attainmentof
    - 50% reduction in the weight of vehicle structure andsubsystems,
    - affordability,and
    - increased use of recyclable/renewablematerials.

An expanded initiative, entitled the FreedomCAR and FUEL initiative, that focuses more on infrastructure issues, has been announced recently.

> "Tonight I am proposing $1.2 billion in research funding so that America can lead the world in developing clean, hydrogen-powered automobiles."
>
> — President George W. Bush, State of the Union Address, January 28, 2003

---

[1] "Renewable" energy generally refers to solar, wind, geothermal, tidal, hydroelectric energy, and biomass.

Technical goals are not yet available, but the general goals are [2]:

- **Lowering the cost of hydrogen:** Currently, hydrogen is four times as expensive to produce as gasoline (when produced from its most affordable source, natural gas). The FreedomCAR and Fuel Initiative seeks to lower that cost enough to make fuel cell cars cost-competitive with conventional gasoline-poweredvehicles by 2010; and to advance the methods of producing hydrogen from renewable resources, nuclear energy, and evencoal.
- **Creating effective hydrogen storage:** Current hydrogen storage systemsare inadequate for use in the wide range of vehicles that consumersdemand.
- **Creating affordable hydrogen fuel cells:** Currently, fuel cells are ten times more expensive than internal combustion engines. The FreedomCAR and FUEL Initiative is working to reduce the cost to affordable levels.

Not everyone was pleased. In the February 2, 2003 issue of the *New York Times*, several groups (Cato Institute, National Taxpayers Union, the Competitive Enterprise Institute, and the Natural Resources Defense Council) listed the "hydrogen car" as a prominent item to cut from the federal budget to reduce the projected shortfall.

GLOSSARY

ICE	Internal combustionengine
NG	Naturalgas
CNG	Compressed naturalgas
SI	Sparkignited
HEV	Hybrid electricvehicle
IC	Internalcombustion
LHV	Lower heatingvalue
CI	Compressionignited
SIDI	Spark ignited directinjection
CIDI	Compression ignited directinjection
FC	Fuelcell
GNF	Graphite nanofiber
SWNT	Single-walled carbonnanotube
MWNT,CNT	Multi-walled carbonnanotube

SYSTEM CONSIDERATIONS

In the April, 2002 issue of *Physics Today* Joan Ogden of the Princeton Environmental Institute discussed the future of hydrogen as a fuel and described the operation of hydrogen-oxygen fuel cells. Ogden stated that practical fuel cells[2] are up to 60% efficient in converting hydrogen energy into electrical energy (which is the FreedomCAR target) although not necessarily at the rated power, significantly higher than the 45% efficiency of using hydrogen in an internal combustion engine. However, these estimates do not include the losses in producing hydrogen from various hydrocarbon sources (Fig. 2). Clearly, hydrogen is not a naturally occurring terrestrial fuel. Rather, it is an energy carrier.

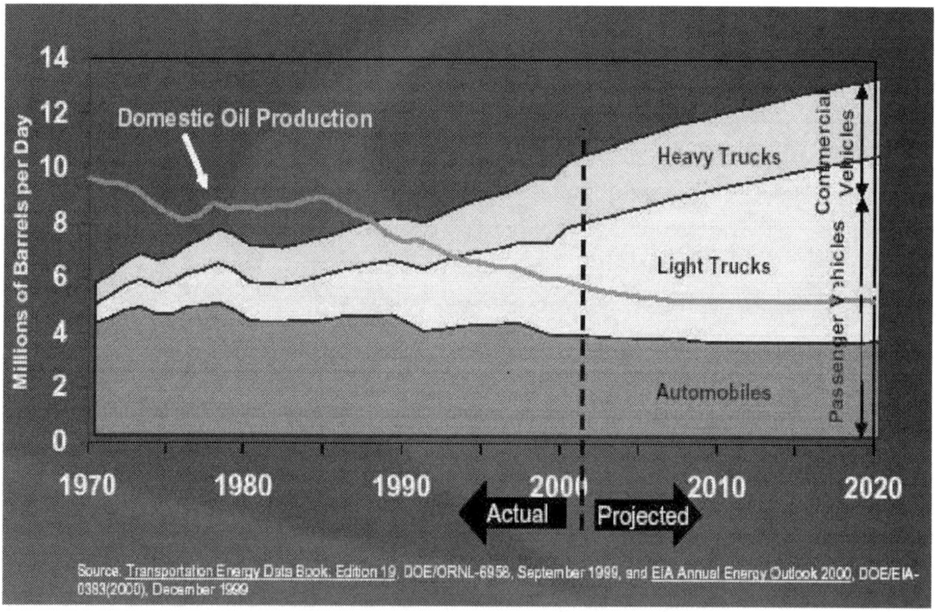

Figure 1: Taken from Ref. [3].

A typical well-to-wheels analysis is shown in Fig. 3. Although the details of the analyses behind Fig. 2 are not readily available, the kind of breakdown of energy losses shown in Fig. 3 underlies each the powertrain options considered.

[2] We refer to proton-exchange membrane fuel cells in this report.

NET ENERGY LOSSES "WELLS TO WHEELS" FOR FUEL CELL VEHICLES

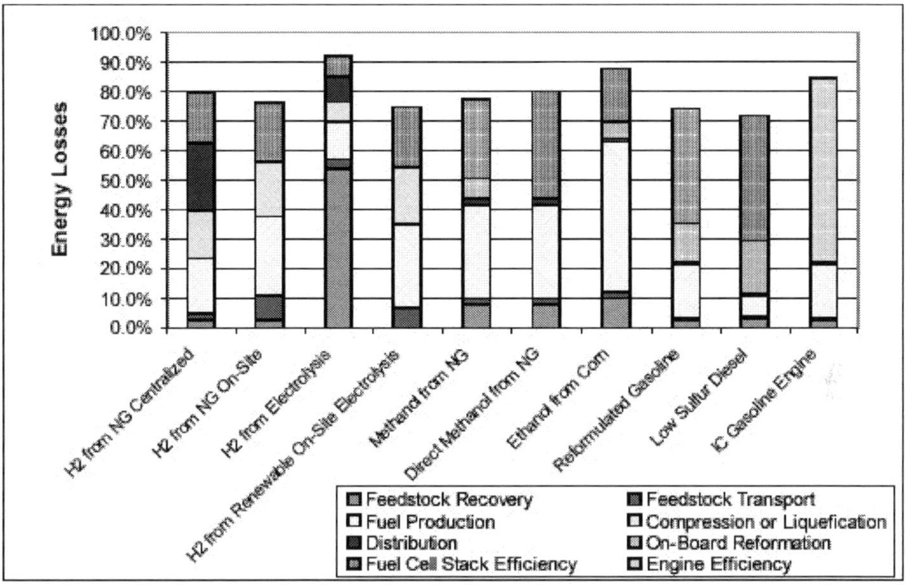

Sources: Albert Sorbey and Associates, 1996; Pembina Institute for Appropriate Development, 2000; Karlhammer, 1997; Moore, 2000; Thomas et al., 1998; and Stodolsky et al., 1999.

Figure 2: Due to way this graph has been drawn, to obtain the energy efficiency of any system, subtract the height of the bar from 100%. From Ref. [4].

For example, the equivalent fuel economy of a compressed natural gas, spark ignited, hybrid electric vehicle (CNG SI/HEV) is 48.6 mpg, whereas a fuel cell vehicle powered by hydrogen derived from methane is projected to get 82.0 mpg, a substantial improvement. However, if viewed from the standpoint of well-to-wheels energy consumed per unit of distance traveled, the difference is more modest: 2867 versus 2368 BTU/mi. If CO_2 is sequestered in the forming of hydrogen, the amount emitted into the atmosphere is only 25 g/mi for the fuel cell vehicle compared to 196 g/mi for the CNG vehicle. This additional benefit favors the fuel cell vehicle. Clearly hydrogen fuel cells do not entirely eliminate CO_2 emissions unless the hydrogen is generated without combusting or reforming hydrocarbon fuels, *e.g.* by electrolysis of water using nuclear, solar or wind power. The emission of CO, NO_x and hydrocarbons associated with the ICE are removed, but may be emitted to some extent in a different location by the chemical plant generating H_2.

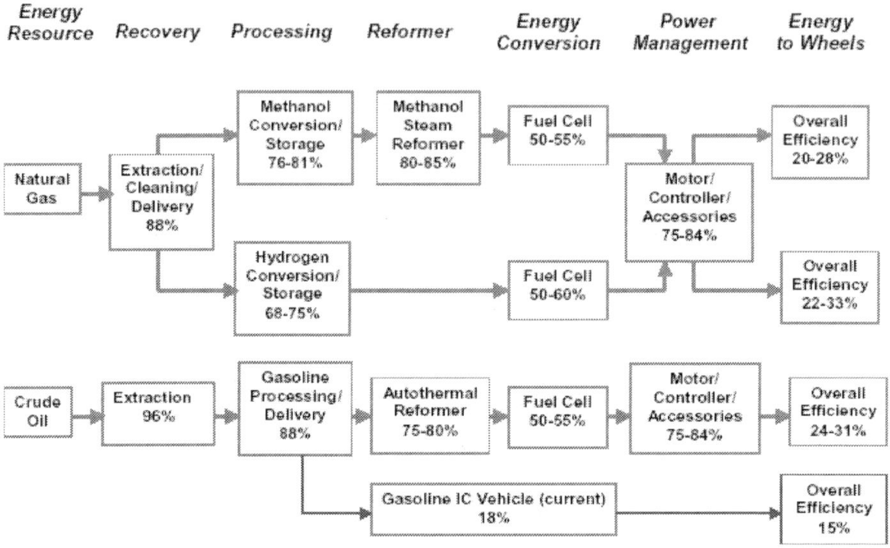

Figure 3: From Ref. [5].

Even if production losses are taken into account, the fuel cell vehicle surpasses the conventional internal combustion engine in efficiency, although the overall efficiency is only about 30% in the best case, less than the 60% x 70% = 42% well-to-wheels efficiency objective of FreedomCAR (Fig. 4), where 60% is the energy-to-wheels goal and 70% is the well-to-pump efficiency.

Further comparison of internal combustion engine and fuel cells for fuel economy and CO_2emissions is shown in Table I.

In a draft report prepared last year for the Office of Transportation Technologies, Energy Efficiency and Renewable Energy, U.S Department of Energy, [6] national lab scientists and others indicated an optimistic mid-term future for fuel cell vehicles (TableII).

Ogden pointed out major obstacles that must be overcome before automotive fuel cell technology can be considered viable. First, today's cost of $1500 to $10 000 per kilowatt of power must come down to the range of $50-100 per kilowatt to be competitive.

According to Ogden, the most expensive component is the membrane electrolyte, typically made of the polymer Nafion. Also, A.D. Little has indicated that the current platinum requirement for a 50-kW system would cost $57/kW, which is higher than the FreedomCAR cost target for the entire fuel cell system. [7][3] [See Appendix C for another

[3] This analysis, however, seems high. By 2000, Pt usage had been reduced to about 0.5 g/kW, [8] which gives a cost of roughly $10/kW for the catalyst.

cost analysis.] Second, a breakthrough in on-board hydrogen storage is required. The currently preferred method is to use a carbon-fiber wrapped compressed-gas cylinder (at a pressure of 34 MPA or 5000 psi, with mass of 32.5 kg, and volume of 186 L for a 500-km range). [See the next section for a discussion of hydrogen storage technology.] An infrastructure to produce and distribute hydrogen economically is the third major problem to be solved. Presently most hydrogen is produced thermochemically (500- 1700 C) in oil refineries and chemical plants by reforming natural gas and other hydrocarbons with steam or oxygen. Unlike petroleum, natural gas supplies are abundant and come mostly from within the United States or are imported from Canada. Production facilities operate at approximately 70% of capacity and the distribution infrastructure has excess capacity [4]. Thus, at least initially, natural gas production and distribution does not appear to be a limiting factor in the availability of hydrogen.

Table 1: Fuel Economy, Energy Use and CO_2Emissions for Alternative Fueled Automobiles [9].

	FUEL ECONOMY Mpg equiv- LHV basis (from GREET model; except fuel cell vehicles and H_2ICE/HEVsfrom DTI)	Well to Wheels Energy Consumption (BTU/mi)	Well to Wheels CO_2emissions (g/mile)
ICENGINE VEIICLES			
Conventional Gasoline SI Engine	*22.4*	6492	514
CNG SI Engine	*20.3*	6702	459
Adv. Diesel CI Engine	*37.0*	4565	378
ICE/IYBRID VEIICLES			
Gasoline SIDI/HEV	*46.9*	3092	252
CNG SI/HEV	*48.6*	2867	196
Ethanol SIDI/HEV	*46.9*	4921	67
H_2SI/HEV	**50.0**	3466 w/o CO_2seq 3580 w/ CO_2seq	234 w/o CO_2seq 41 w/ CO_2seq
Diesel CIDI/HEV	*56.8*	2487	208

FUEL CELL VEIICLES			
Gasoline (probable) (best)	**38.0** **49.4**	3819 2938	304 234
Methanol (probable) (best)	**56.0** **64.2**	3212 2802	199 174
Hydrogen (from natural gas with steamreforming, pipeline deliveryand compression to 5000 psi foronboard storage	**82.0**	2368 w/o CO_2seq 2446 w/CO_2seq	143 w/o CO_2seq 25 w/CO_2seq.

Fuel economies shown in *italics* are from the GREET model [10]. Fuel economies in boldface are from the DTI model [11]. CNG SI/HEV vehicle fuel economies have been scaled from GREET results.

Table 2: Source: Ref. [6].

Vehicle System	Fuel Economy Improvement Potential	Criteria Emissions	Years to Mass Market Introduction	Current Incremental Cost	Other Issues
Enhanced Conventional	Moderate (50%)	Continued though reduced	Very near term (0-5 y)	Minimal (5%)	High consumer acceptance, continued petroleum dependence
Hybrid	Substantial (100-200%)	Some zero emission range possible	Near term (2-7 y)	Substantial (10-20%)	Grade climbing ability or towing capacity may be reduced
Fuel Cell	Very High (150-300%)	Low to zero tailpipe and total	Mid term (7-12 y)	Very high (>20%)	Potential petroleum independence
Battery-Electric	Very High (300%)	Zero tailpipe	Near term (2-7 y)	Very high (>20%)	Energy storage, range concerns, low petroleum use

A different approach that does not rely on hydrocarbons has been analyzed by C. W. Forsberg and K. L. Peddicord [12]. They discussed the economics of H_2production using nuclear energy to provide the energy for electrolysis of water and concluded "The technology has the potential for economic production of H_2." Likewise, hydrogen production from renewable sources such as wind power could be interesting, but has not been analyzed here. Fortunately, the distribution of H_2may not be as daunting as one might think. Fosberg, in a private communication, noted the existence of several hydrogen pipelines in Europe, the United States and Japan. However, natural gas lines would have to be retrofitted with new valves and compressors before hydrogen couldbe transported through them.

Initially the auto industry felt that sufficient hydrogen fuel would not be available quickly enough, so engineers pursued a path that required on-board reforming of gasoline or methanol. DaimlerChrysler demonstrated an example in October 2000. The Jeep Commander 2 (similar to the Jeep Grand Cherokee sport utility vehicle) reformed pure, electronic-grade methanol to power two Ballard fuel stacks. Although DaimlerChrysler demonstrated 23.5 mpg fuel efficiency (almost twice that of a comparable gasoline vehicle) with acceptable performance and acceleration, they found that fuel reforming must be improved because the cold-start time was unacceptable [13].

Comparison of Energy Efficiencies for Fuel Cells, Internal Combustion and Iybrid Vehicles

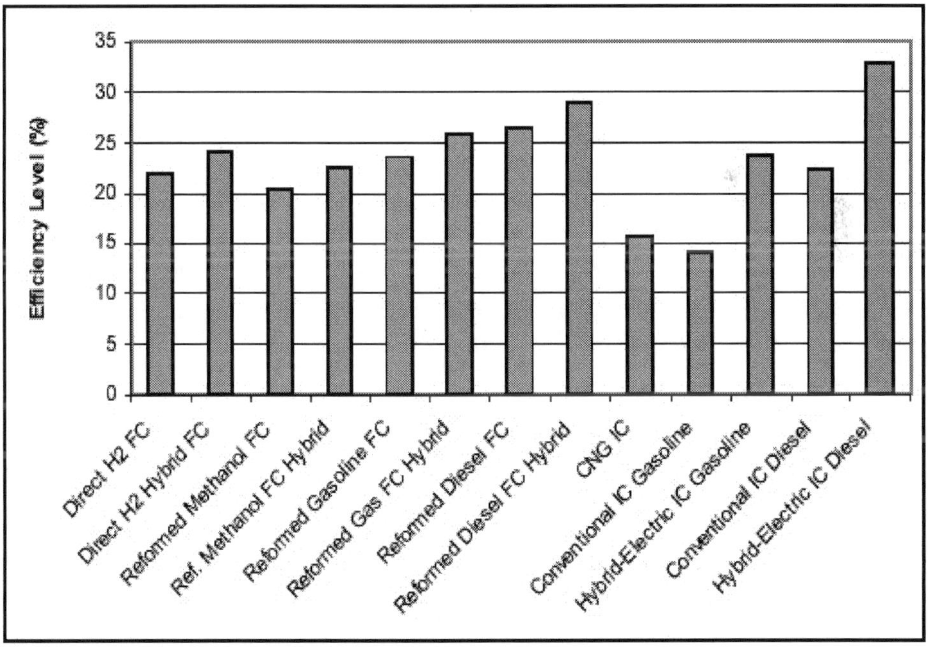

Sources: Stodolsky et al., 1999; Albert Sorbey and Associates, 1996; Pembina Institute for Appropriate Development, 2000; and Thomas et al., 1998.

Fig. 4 From Ref [4].

On May 1, 2002, according to a press release, General Motors demonstrated the world's first drivable fuel cell vehicle (a Chevrolet

S-10 fuel cell pickup) that extracts hydrogen from gasoline. "This vehicle and the reforming technology in it move us closer to a hydrogen economy," said Larry Burns, GM's Vice President of Research and Development, and Planning. The fuel cell pickup was equipped with a fuel processor that reformed low-sulfur gasoline. When linked with a fuel cell stack, GM said the vehicle could achieve up to 40 percent overall energy efficiency, which is a 50 percent improvement over a conventional internal combustion engine. For further information, see Burns *et al.* [14].

Ron Sims, Ford Motor Co. research engineer (retired) and consultant to ORNL, feels that gasoline reformers (and presumably methanol as well) on-board the vehicle are no longer viable because they are too costly and too complex. Obviously, the reformer adds another chemical plant to the vehicle-an undesirable feature. However, stationary reformers at gas stations might makesense.

Sims thinks it will take 10-15 y for commercialization of fuel cell vehicles, 20 y before internal combustion engine sales will notice the impact of fuel cells. On the other hand, Larry Burns and other GM executives have publicly stated, "By the end of this decade, you can expect to see affordable, profitable fuel cell vehicles on the road."

By 2000, Ford had built a hydrogen refueling station at the Engineering and Research Center in Dearborn, Michigan (Fig. 5) and had developed a hydrogen (no reforming) fuel cell vehicle with on-board storage of compressed gas (Fig. 6).

According to Ref. [14] Toyota is preparing a fuel cell hybrid vehicle, called FCHV-4, for production. Two vehicles have been delivered to the University of California for research purposes [15]. The Toyota vehicle uses compressed hydrogen gas, as does the Honda FCX, currently being tested in California. The 2003 Honda FCX has just been certified in the US as a zero emission vehicle. [16] General Motors exhibited a new prototype fuel cell hybrid named "Hy-wire" at the Paris Motor Show in September 2002. Although this vehicle has a top speed of 100 mph, it has a range of only 100 miles, far short of the acceptable driving range of 300 miles. In the May 2002 press release Larry Burns, while still maintaining that GM will produce affordable, customer-friendly fuel cell vehicles by 2010, believes GM will only "sell them profitably and in large numbersby 2020." Clearly the timetable is rather long, consistent with the opinion of Ron Sims. For more information on prototype fuel cell vehicles, see the article by Jost [17] and Ref. [18].

Other engineering issues that all manufacturers face, although seemly mundane, are nonetheless challenging:

- Cold weather operation

- Packaging
- Reliability
- Safety (including public acceptance of H_2fuel) [See section on safetybelow.]
- Manufacturability

In an interview we had with Prof. J. Schwank, U. of Michigan Chemical Engineering Dept., he suggested that the hydrogen internal combustion engine is potentially a better bet than fuel cells because hydrogen does not then have to be as highly purified. The ICE can take as much as 10% CO in the fuel, which would be lethal for the fuel cell.

Purification of the hydrogen stream to remove CO is a significant cost. In Table I, the fuel economy of the H_2SI/HEV (spark-ignited, hybrid-electric vehicle) is estimated to be 50 mpg with about 3500 BTU/mi (2.3 MJ/km) well-to-wheels energy consumption compared to 2400 BTU/mi (1.6 MJ/km) for fuelcells.

Figure 5: Source: Ref. [8].

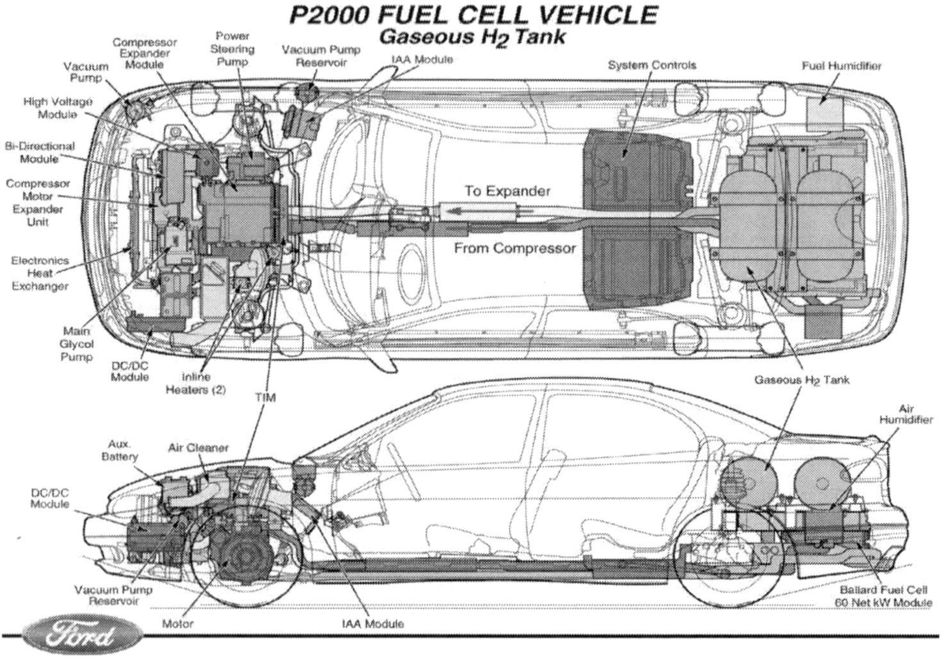

Figure 6: Source: Ref. [8].

Schwank still feels that fuel processors, which reform hydrocarbon fuels to produce hydrogen, are important and they need to be highly flexible because of the difference in fuels by season and region of the country. For military applications, the processor must be able to run on the highly developed (and unlikely to be changed!) JP8 fuel.

The Army estimates it costs $600/gal to carry fuel to the battlefield, so reducing consumption is essential. Instead of powering vehicles, fuel cells could be used as auxiliary units to run on-board electronics and communications equipment. Quiet operation to generate electricity for stationary field applications is just as important to the military as higher fuel efficiency. The Army is presently letting contracts for demonstration fuel cell powered trucks [19]. The impact that military funding might have for fuel cell development is huge.

In addition to USCAR, which is the automotive industry consortium that has responsibility for implementing many of the FreedomCAR and FUEL programs, the California Fuel Cell Partnership [20] is actively testing, promoting, and demonstrating fuel cell technology. This partnership of automotive manufacturers, energy companies, fuel cell companies, and government intends to evaluate as many as 60 vehicles in real- world conditions.

Recently, MIT issued a detailed report comparing life-cycle energy costs that includes energy expended in manufacturing and final disposal as well as for fuel consumption on the road and during production and distribution. Weiss *et al.* [21] made reasonable assumptions about the future evolution of internal combustion engines andhybrid-electric vehicles in addition to predicting the pace of development for fuel cell vehicles. On the basis of life-cycle energy consumption, the differences between fuel cell and hybrid- electric vehicles (mid-size) are small compared to the uncertainties in thepredictions.

The manufacturing energy requirements do not differ substantially, but the predicted fuel economy of a 2020 diesel ICE hybrid electric is high enough and the fuel cycle energy low enough to offset much of the advantage of fuel cells. [See Appendix D.] Their study pertains to the next twenty years and to energy derived from petroleum or naturalgas. Similar conclusions were found for green house gas emissions.

HYDROGEN STORAGE TECHNOLOGIES

Introduction

In the development of fuel cell vehicles, hydrogen storage is "the biggest remaining research problem" according to the January 2003 Office of Technology Policy report, *Fuel Cell Vehicles: Race to a New Automotive Future* [22]. Current hydrogen storage systems are inadequate to meet the needs of consumers in a fuel cell vehicle. The OTP report continues, "Hydrogen's low energy-density makes it difficult to store enough on board a vehicle to achieve sufficient vehicle range without the storage container being too large or too heavy."

Existing and proposed technologies for hydrogen storage include

1. physical storage: pressurized tanks for gaseous hydrogen and pressurized cryotanks for liquid hydrogen;
2. reversible hydrogen uptake in various metal-based compounds including hydrides, nitrides, and imides;
3. chemical storage in irreversible hydrogen carriers such as methanol;
4. cryoadsorption with activated carbon as the most common adsorbent; and
5. advanced carbon materials absorption, including carbon nanotubes, alkali-doped carbon nanotubes, and graphite nanofibers. The U.S. Department of Energy report, *A National Vision for America's Transition to a Hydrogen Economy-To*

2030 and Beyond, projects that pressurized tanks will be the predominant hydrogen storage technology until about 2015, to be supplanted by hydride storage into the early 2020s, then other solid state storage technologies [23]. They see storage technologies maturing sufficiently for mass production in the2020s.

The Department of Energy timeline for development of storage systems projects that high pressure and cryogenic storage will be demonstrated in 2002-3, cost-effective hydride storage systems in 2003-6, and carbon-based storage systems in 2006-11 [24].

Goals for hydrogen storage systems for 2010 that were established in the FreedomCAR initiative [2] include

- available capacity of 6 wt%hydrogen
- specific energy of 2kWh/kg
- energy density of 1.1kWh/L
- cost ≤ $5/kWh or $1.25/gal (gas equiv.) in CY2001dollars

Research into hydrogen storage technologies is still in its infancy, as reflected in the very low level of patenting in this area: 14 patents in 2001, and fewer in most previous recent years [25]. Patent data indicate that the United States is the "leader" in this research area, while "US-based Energy Conversion Devices and Canadian organization Hydro-Quebec and McGill University is where the action appears to be located" [26]. Likewise, only a few papers on hydrogen storage were presented at the recent (2003) March APS Meeting.

According to John M. Decicco [27], as quoted in Ref. [22], p. 24, hydrogen storage faces deployment barriers that are

- fundamental (basic research stillneeded)
- developmental (engineering R&D required for practicaldesigns)
- maturational (mass-production commitmentspremature)
- experience-limited (costs higher than long-run potential due to lack of production experience)
- infrastructural (lack of appropriate fuel or servicefacilities)

This report summarizes the current status of hydrogen storage technologies and directions of current and needed research in this area.

Pressurized Tank Storage

Pressurized tanks of adequate strength, including impact resistance for safety in collisions, have been made of carbon-fiber wrapped cylinders. Compressed gas storage in such tanks has been demonstrated at a pressure of 34 MPa (5,000 psi) with a mass of 32.5 kg and volume of 186 L, sufficient for a 500-km range. Note, however, that this tank volume is about 90% of a 55-gallon drum, rather large for individual automobiles. So while the 6 wt% goal can be achieved, tank volume is problematic. Pressures of 70 MPa (10,000 psi) have been reached, and in 2002 Germany certified Quantum Technology's 10,000 psi on-board storage tank [24]. A footnote in the OTP report cited above [22] says, "The Toyota and Honda vehicles available for lease in late 2002 use hydrogen stored in high-pressure containers [28]. However, their range will be less than optimal because hydrogen's low density does not permit a sufficient amount to be stored (unlike CNG [compressed natural gas], which has a higher energy density for the samevolume)."

Low temperature storage of liquid hydrogen does not appear to be suitable for normal vehicle use, although research on this possibility is being conducted at a low level by several automobile manufacturers [24]. Furthermore, "a liquid hydrogen storage system loses up to 1% a day by boiling and up to 30% during filling, as well as requiring insulation to keep the hydrogen at 20 K." [29]

Hydrogen Uptake in Metal-Based Compounds

Metal hydridation can be used to store hydrogen above room temperature and below 3 or 4 MPa. However, the metals introduce too much additional weight for most vehicle uses. They are also expensive [30].

Recent work by P. Chen, et al. has shown that lithium nitride can reversibly take up large amounts of hydrogen [31]. This material takes up hydrogen rapidly in the temperature range 170-210°C, and achieved 9.3 wt% uptake when the sample was held at 255°C for 30 minutes. Under high vacuum (10^{-9}MPa, 10^{-5}mbar, or 10^{-5}torr) about two-thirds of the hydrogen was released at temperatures below 200°C. The remaining third of the stored hydrogen required temperatures above 320°C for release. The hydrogen was taken up as lithium imide ($LiNH_2$) and lithium hydride (LiH). These researchers suggest that related metal-N-H systems should be investigated to find a hydrogen storage system that works at more practical temperatures and pressures.

Cryoadsorption Hydrogen Storage

While having potential weight and volume advantages, cryoadsorption with activated carbon as adsorbant requires liquid nitrogen temperatures and 2 MPa (300 psi) to hold the physically adsorbed hydrogen [32]. It does not appear to be suitable for vehicle use.

Carbon Nanotube and Related Storage Technologies

The status of hydrogen storage in advanced carbon materials is still unclear. In this subsection, we review briefly the status of carbon nanotube storage, both single-walled and double-walled, and graphite nanofiber stack storage. Other carbon-based storage technologies that have been proposed include alkali-doped graphite, fullerenes, and activated carbon.

High surface area and abundant pore volume in the nanostructured materials make these especially attractive as potential absorption storage materials. Some early work gave tantalizing results for hydrogen storage in carbon nanotubes. Ogden reported various conflicting, some excessively optimistic, results [1]. A query by this subcommittee to Prof. Mildred Dresselhaus of MIT about the achievable wt% (6.5 wt% has been suggested) brought this response [33]:

1. It is hard to say what is a reliable estimate of the hydrogen uptake number because of the differences in the reported levels by different groups, presumably doing similar measurements. The reasons for the different results between groups are notunderstood.
2. The 6.5% value is not yet achievable in myopinion.
3. The problem seems to be hard to me, arguing from a theoretical standpoint. However I would not discount the possibility of a break through that might change the situation dramatically. So far it doesn't seem to me that there is yet much available carefully controlled work.

A 2001 review of carbon nanostructure storage research, sponsored by the German Federal Ministry for Education and Research (BMBF), found that follow-up work "has been unable to reproduce any of the high-capacity results." [29] They concluded "In view of today's knowledge, it is unlikely that carbon nanostructures can store the required amount of hydrogen. In any case, this calls into doubt whether carbon nanostructures would have any advantage over high-pressure tank storage."

The German study [29] presented the following summary figure on the capacity of current and future hydrogen storage systems:

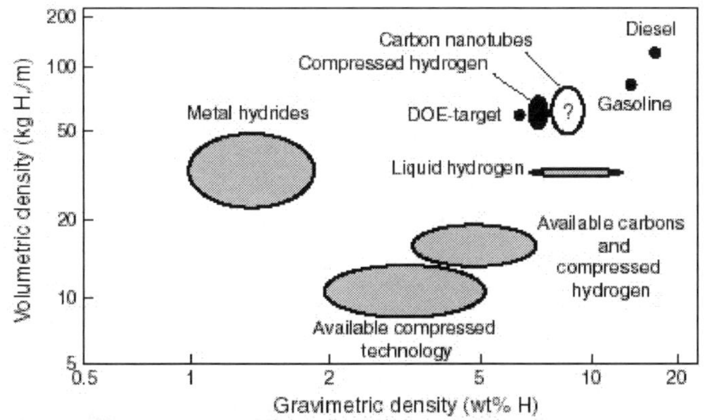

Figure 1. Summary of how current and future hydrogen storage systems relate with respect to gravimetric and volumetric densities.[3]

Figure 7: Summary of how current and future hydrogen storage systems relate with respect to gravimetric and volumetric densities.

The considerable doubt that has been cast upon carbon nanotube storage capacity makes it highly likely that their position on this chart will have to be moved to a much less favorable point.

A summary of results was given by Ding et al. [34] and reproduced in Ref. 29.

Among all the reports on carbon nanotube storage capacities, one finds that the reproducible results are invariably lower than the earlier encouraging findings. So values of 1-4 wt% are the best that can be depended upon with present data. These are not sufficient to meet the benchmark of 6 wt%.

A review and research report published in 2002 considered hydrogen storage by graphite, graphite nanofibers (GNFs), and single-walled carbon nanotubes [35]. These authors concluded "These investigations show a reversible hydrogen storage only for SWNTs and in addition indicate that an opening of the SWNTs is essential to reach high storage capacities." Ref. 29 points out that results on GNFs have been mixed and contradictory, with most high capacity results being explained subsequently "by the presence of water vapor, which expanded the spacing between graphite layers (typically -3.4 A) to accept multiple layers ofhydrogen." [36] Nevertheless, relatively high storage measurements for GNFs (-10-13 wt%) have been reported for GNFs grown by thermal cracking [37] and for vapor-grown GNFs (with some conflicting later results by the same group) [38,39].

Further research is clearly needed before reliable conclusions can be drawn about GNFs.

Table 3: Summary of reported results achieved for hydrogen storage in carbon nanostructures. SWNT = single-walled carbon nanotubes, MWNT and CNT = multi-walled carbon nanotubes. Table references are appended following the endnotes to this report. (From Ding et al., Ref. 34)

Material	Temperature	Pressure, MPa	Max. wt% of H_2	Reference
SWNT 100%	133 K (–140°C)	0.04	5–10 (prediction)	**Dillon 1997**
SWNT High purity	Ambient	0.067	~3.5–4.5	**Dillon 1999**
SWNT ~50%	300 K (27°C)	10.1	~4.2	**Liu 1999**
SWNT High purity	80 K (–193°C)	~7	8.25	**Ye 1999**
MWNT	~300–700 K (27–427°C)	Ambient	0.25	**Wu 2000**
SWNT-$TiAl_{0.1}V_{0.04}$	Ambient	0.067	~7	**Dillon 2000**
SWNT-Ti-6Al-4V	Ambient	0.08	1.47	**Hirscher 2001**
SWNT-Fe	Ambient	0.08	<0.005	**Hirscher 2001**
SWNT ball milled in Ar	Ambient	0.08	<0.1	**Hirscher 2002**
SWNT ball milled in D_2	Ambient	0.9	1.0	**Hirscher 2002**
Li-CNT	473–673 K (200–400°C)	0.1	20	**Chen 1999**
Li-CNT	<313 K (40°C)	0.1	14	**Chen 1999**
Li-CNT (wet H_2)	473–673 K (200–400°C)	0.1	12	**Yang 2000**
Li-CNT (dry H_2)	473–673 K (200–400°C)	0.1	2.5	**Yang 2000**
K-CNT (wet H_2)	<313 K (40°C)	0.1	21	**Yang 2000**
K-CNT (dry H_2)	<313 K (40°C)	0.1	1.8	**Yang 2000**
Li-CNT	473–663 K (200–400°C)	0.1	0.72–4.2	**Pinkerton 2000**

Early work that found up to 20 wt% hydrogen uptake by lithium-doped carbon nanotubes and up to 14 wt% by potassium-doped carbon nanotubes [40] was not reproducible, and it was later shown that most of the weight gain was due to water rather than hydrogen [41].

A "Review of theoretical calculations of hydrogen storage in carbon-based materials" was carried out by Meregalli and Parrinello in 2000 [42]. They drew two major conclusions: (1) "The reported calculations indicated a hydrogen uptake smaller than the more optimistic experimental results." (2) "Furthermore the calculations suggest that a variety of complex chemical processes could accompany hydrogen storage and release." In particular, they find both physisorption and chemisorption contributing to hydrogen uptake. They found potential for up to 12.5-wt% storage in GNFs, with most results below the 6-wt% target. For SWNTs, calculations gave a theoretically possiblemaximum of 14-wt% and for MWNTs, 7.7 wt%, relatively encouraging if such numbers can be achieved experimentally. However, the amount available for release could be small compared the amountstored.

Section Summary

The only proven system for hydrogen storage today that is practical for fuel cell vehicles is compressed gas high pressure tank storage. While this technology has the disadvantages of limited energy density and possibly high weight for the tank, it has been shown to workable with up to 70 GPa (10,000 psi) on-board storage tanks.

Of the other hydrogen storage systems, advanced carbon materials have been especially intriguing possibilities. However, very mixed experimental results have left us to conclude that these materials are far from proven to have adequate storage capacity.

Further research is ongoing, but a breakthrough is needed to provide a foundation for confidence that carbon nanotubes or related materials will be able to satisfy storage requirements.

Finally, new ideas in hydridation of metal-N-H systems are sufficiently interesting that they are being pursued and may lead to the development of practical storage systems.

HYDROGEN FUEL STORAGE SAFETY

Hydrogen has a reputation for being explosive and therefore raises concerns about the safety of carrying a substantial quantity of H_2in a vehicle fuel tank. However, because $H_2$is the lightest gas, it has a tendency to diffuse away quickly in case its container is breached and consequently may represent less of a hazard than gasoline.

The simplest way to carry hydrogen fuel in a car or other vehicle is as a high-pressure gas 3-10 kpsi (21-69 MPa) in metal or composite-reinforced (fiberglass, carbon fiber, Kevlar) tanks. This is similar to the way compressed natural gas (CNG) vehicles operate.

There is an interesting report on $H_2$for energy use [43] by the Norwegian environmental organization Bellona with useful safety information in Chapter 5. These authors conclude that "hydrogen is no more or less dangerous than any other energy carrier and furthermore that hydrogen has properties that in certain areas make it safer than other energy carriers: it is not poisonous, and has the ability to dissipate quickly into the atmosphere because of its light weight compared to air." They describe tests by Lockheed and Arthur D. Little that indicated that $H_2$is, if anything, safer than gasoline or jetfuel.

There are a number of references to crash testing by BMW [43, 44] that say BMW demonstrated the safety of H_2fuel for cars.

Regarding the Hindenburg accident, a recent study [45] concluded that the paint on the dirigible skin was extremely flammable and was the true cause of the disaster. Of the 61 crew plus 36 passengers (total 97), 35 died, plus one person on the ground. Of the 35, 27 jumped when the Hindenburg was in the air and 8 were killed by burning dieselfuel [43,45]. There was no explosion. "The fire of the hydrogen from the gas cells lasted only less than one minute, and there is no evidence that anybody was directly hurt by it" [45].

Dr. Michael Swain at the University of Miami set fire to two cars, one carrying gasoline and the other hydrogen [46, 47]. The gasoline car had a 1/16" puncture in a fuel line. The hydrogen car had a leaking hydrogen connector. The gasoline-filled car was completely destroyed in an intense fire and the hydrogen car was essentially undamaged.

CONCLUSIONS

The efficiency of fuel cells in practical driving conditions can be 50-60%, but the overall efficiency (wells-to-wheels) is no more than 30% when production of H_2 from hydrocarbon resources is taken into account, significantly less than the FreedomCAR goal (60%×70% = 42%).[4] In comparison, the overall efficiency for the conventional internal combustion (IC) engine is about 15%. Hybrid-electric gasoline ICs are expected to reach almost 25%. On-site H_2production by electrolysis from renewable energy sources is about 50% efficient, giving an overall fuel cell vehicle efficiency of 25%.

Reforming natural gas is the most practical source of H_2at present. Nuclear power production of hydrogen, however, may prove practical in the long term.

For fuel cell vehicles to be economically viable, the cost of fuel cell stacks must be reduced to -\$100/kW compared to the \$1500/kW achieved thus far. Advances in catalyst (Pt) technology have significantly reduced expensive Pt usage. The FreedomCAR cost target is \$45/kW.

Hydrogen storage is still a major research problem. While progress has been made, current systems are inadequate or marginal. On-board storage of sufficient H_2 most likely will be as compressed gas. Demonstration vehicles have used 5000-psi carbon-fiber wrapped

[4] Hopefully, the sources of energy for H_2production will eventually be renewable.

tanks, and tanks have been certified recently at pressures of 10,000 psi. Liquid hydrogen and cryoadsorption storage are almost certainly not practical for vehicle use.

Metal hydridation and metal-N-H systems show some promise, but much research is still required for these systems. Advanced carbon materials are intriguing, but contradictory research results, with relatively low reproducible storage capacities, have been discouraging and much remains to be done. Unresolved research questions, such as hydrogen storage, make the fuel cell vehicle approach a long-term issue. It is not a short-term solution to energy or emissions problems.

For the next twenty years, vehicles powered by the hybrid-electric internal combustion engine could well prove to be just as beneficial, from a life-cycle perspective, as fuel cell vehicles. A recent MIT study [21] finds that fuel cell vehicles do not have significant advantages (in terms of energy consumption or CO_2 emissions) over future vehicles powered by hybrid-electric, internal combustion engines.

Obviously no widespread infrastructure exists, but distribution might not be as daunting as first thought. Several pipelines for H_2 transmission already exist and natural gas lines could be retrofitted for these purposes. Transportation by tanker truck over short distances is well established. The FreedomFUEL initiative recently announced by President Bush is intended to address some of the infrastructure and cost issues. Realistic estimates of when substantial market penetration of fuel cell vehicles will occur (at least 15 years) allow plenty of time to develop an infrastructure. Scenarios of how a hydrogen society might evolve have been explored by Burns *et al.* [14] and Chinworth [48].

The motivation for the FreedomCAR and FUEL initiatives is to reduce US petroleum imports, greenhouse gas emissions, and atmospheric pollutants. This study shows that hydrogen fuel cell vehicles as well as advanced internal combustion hybrid electric vehicles can accomplish significant reductions in energy usage by the transportation sector with the concomitant reduction of CO_2emissions. Roughly, a factor of two reduction in life-cycle energy (and CO_2) per unit of distance traveled relative to today's vehicles is possible. The emissions of atmospheric pollutants such as CO and NO_xdepend strongly on the source of hydrogen, however. Only generation of hydrogen by renewable energy sources or by nuclear power truly reduces emissions of pollutants, although the point of origin may be away from urban areas. The FreedomCAR initiative includes more than just fuel cells, which is fortunate because it is not clear what technology may be the best for the future.

Nonetheless, there was optimism at a recent DOE workshop [53] that,despite daunting challenges, many good ideas remain to be pursued. Research carried out for hydrogen initiatives could well improve long-term energysecurity.

Acknowledments

The committee acknowledges input from Joan Ogden (via David Chock). We wish to thank Ron Sims, Daniel Sperling, Johannes Schwank, Charles Forsberg, and Mildred Dresselhaus for their contributions. The authors are indebted to Barbara Levi for a critical reading of an earlier draft and for many helpful editorial suggestions.

General Disclaimer

This POPA E&E Subcommittee report represents only the views and estimates of individual members and contributors and not necessarily of their organizations.

REFERENCES

"Hydrogen: The Fuel of the Future?" Joan M. Ogden, *Physics Today*, April 2002,pp. 69-75.

http://www.cartech.doe.gov/freedomcar/technical-goals.html;http://www.eere.energy.gov/hydrogenfuel/

"Progress Towards Environmentally Friendly Vehicles," American Physical Society, March 19, 2002 by Bob Culver, Executive Director USCAR.

"Future Wheels: Interviews with 44 Global Experts On the Future of Fuel Cells for Transportation And Fuel Cell Infrastructure AND A Fuel Cell Primer," Northeast Advanced Vehicle Consortium, M.J. Bradley and Associates, November2000.

Fuel Cell Handbook (Fifth Edition), EG&G Services, Parsons, Inc.,Science Applications International Corporation, October 2000. Available to the public from the National Technical Information Service, U.S. Department of Commerce, 5285Port Royal Road, Springfield, VA 22161; phone orders accepted at (703)487-4650.

Future U.S. Highway Energy Use: A Fifty Year Perspective, Dept. of Energy, May, 2001.

USCAR web site,www.USCAR.org.

"ABCs of Automotive Fuel Cells," Ron Sims, PowerPoint Presentation provided by theauthor.

Private communication from Joan Ogden to David Chock, Ford Motor Co., November 13,2001.

Wang, M.Q., 1999: GREET 1.5 - Transportation Fuel-Cycle Model, **1**, *Methodology, Development, Use and Results*, Center for Transportation

Research, Argonne National Laboratory, prepared for the United States Department of Energy Office of. Transportation Technologies, Report No. ANL/ESD-40,December.

Thomas, C.E., B.D. James, F.D. Lomax, and I.F. Kuhn, 1998: *Societal Impacts of Fuel Options for Fuel Cell Vehicles*, Society of Automotive Engineers Technical Paper No. 982496, presented at the SAE Fall Fuels and Lubricants Meeting and Exposition, San Francisco, CA, October 19-22; Thomas, C.E., B.D.James, F.D. Lomax, and I.F. Kuhn, 1998: Draft Final Report, *Integrated Analysis of Hydrogen Passenger Vehicle Transportation Pathways*, report to the National Renewable Energy Laboratory, U.S. Department of Energy, Golden, CO, Under Subcontract No. AXE-6-16685-01,March.

"WHY HYDROGEN PRODUCTION FROM NUCLEAR POWER? (Is Nuclear Energy Compatible with Hydrogen Production?)," C. W. Forsberg and K. L. Peddicord, prepared for presentation at the American Institute of Chemical Engineers' Spring National Meeting, New Orleans, LA, March 12, 2002; "Hydrogen, nuclear energy, and the advanced high-temperature reactor," Charles W. Forsberg, *International Journal of Hydrogen Energy* (in press). Dr. Forsberg remarks, "One of the big attractions for the hydrogen economy worth noting is the ability to minimize pollution. It is politically, technically, and economically easier to control pollution at a thousand industrial sites producing hydrogen than in 100 million vehicles. Hydrogen is clean at the point of use." (private communication, May 3, 2003).

"Jeep Commander 2," *Automotive Engineering International*, July 2002, pp.65-72.

"Hydrogen fuel-cell cars could be the catalyst for a clearer tomorrow," Lawrence D. Burns, J. Byron McCormick and Christopher E. Borroni-Bird, *Scientific American*, October, 2000, pp.65-73.

"Then there were two,' Jack Yamaguchi, *Automotive Engineering International*, August, 2002,pp.17-22.

"Toyota's lease on fuel cells," *Automotive Engineering International*, Jan. 2003,p.43

http://www.epa.gov/newsroom/headline_021103.htm

"Fuel-cell commercialization," Kevin Jost, *Automotive Engineering International*, September, 2002, pp.26-33.

http://www.fuelcelltoday.com/FuelCellToday/EducationCentre/EducationCentreExternal/EduCentreDisplay/0,1748,FCImages,00.html

Forbes, **171**, No. 2, Jan. 20,2003, p.46; Reuters News Service, Mar. 4,2003. www.fuelcellpartnership.org

"Comparative Assessment of Fuel Cell Cars," M. A. Weiss. J. B.Heywood, A. Schafer, and V. K. Natarajan, February 2003, MIT LFEE 2003-001 RP, http://lfee.mit.edu/publications/.

Office of Technology Policy, *Fuel Cell Vehicles: Race to a New Automotive Future* (Washington, DC, January 2003), p. 23.

U.S. Department of Energy, *A National Vision for America's Transition to a Hydrogen Economy-To 2030 and Beyond* (Washington, DC, February 2002), p. iv.

Jerald A. Cole, "Overview of the hydrogen-powered economy - today and beyond," California Hydrogen Business Council, presentation to AssociationofEnergyEngineers, Southern California Chapter, 14 March 2002,http://www.ch2bc.org.

Michael B. Albert, Margaret Cheney, Patrick Thomas, Peter Kroll, and Phyllis Genther Yoshida, *The U.S. Competitive Position in Advanced Automotive-Technologies* (OTP: Washington, DC, 2002), p.24. 26. Ref. 25, p. 29.

John M. Decicco, *Fuel Cell Vehicles: Technology, Market, and Policy Issues*, SAE Research Report, 2001, p.x.

Toyota Motor Company, "Toyota Advances Marketing of Fuel Cell-Electric Hybrid SUV," press release, July 1, 2002; and American Honda Motor Co., Inc., "Honda Fuel Cell Vehicle First to Receive Certification," press release, July 24,2002.

Kaylene Atkinson, Siegmar Roth, Michael Hirscher, and Werner Grunwald, "Carbon nanostructures: An efficient hydrogen storage medium for fuel cells?" *Fuel Cells Bulletin* **4**(38):9-12 (November 2001).

See G. C. Carter and F. L. Carter, *Metal-Hydrogen Systems*, T. Nejat Veziroglu, ed. (Pergamon, Oxford, 1981), Ch. 7 and H. Buchner, P. Pelloux-Gervais, M. Muller, F. Grafwallner, and P. Luger, *Hydrogen and Other Alternative Fuels for Air and Ground Transportation*, H. W. Pohl, ed. (Wiley, Chichester, UK, 1995), Chs.7-11.

Ping Chen, Zhitao Xiong, Jizhong Luo, Jianyi Lin, and Kuang Lee Tan, "Interaction of hydrogen with metal nitrides and imides," *Nature* **420**:302(2002).

See S. Hynek, W. Fuller, and J. Bentley, *Int. J. Hydrogen Energy* **22**:601 (1997) and J. Nitsch, W. Peschka, W. Schnurnberger, M. Fischer, H. Eichert in *Hydrogen as an Energy Carrier*, C. Winter and J. Nitsch, eds. (Springer-Verlag, Berlin, 1988), part B. Personal communication to L. Craig Davis, 23 May2002.

R. G. Ding, G. Q. Lu, Z. F. Yan, and M. A. Wilson, "Recent advances in the preparation and utilization of carbon nanotubes for hydrogen storage," *J. of Nanoscience & Nanotechnology* **1**:7-30 (2001).

M. Hirscher, M. Becher, M. Haluska, A. Quintel, V. Skakalova, Y.-M. Choi, U. Dettlaff-Weglikowska, S. Roth, I. Stepanek, P. Bernier, A. Leonhardt, and J. Fink, "Hydrogen storage in carbon nanostructures," *J. of Alloys & Compounds* **330-332**:654- 658(2002).

C. Park, P. E. Anderson, A. Chambers, C. D. Tan, R. Hidalgo, and N. M. Rodriguez, "Further studies of the interaction of hydrogen with graphite nanofibers," *J. Phys. Chem. B* **103**:10572-10581 (1999).

B. K Gupta and O. N. Srivastava, "Synthesis and hydrogenation behaviour of graphitic nanofibres," *Int. J. Hydrogen Energy* **25**:825-830(2000).

Y.-Y. Fan, B. Liao, M. Liu, Y.-L. Wei, M.-Q. Liu, and H.-M. Cheng, "Hydrogen uptake in vapor grown carbon nanofibers," *Carbon* **37**:1649-1652(1999)

H. M. Cheng, C. Liu, Y. Y. Fan, F. Li, G. Su, H. T. Cong, L. L. He, M.Liu, *Zeitschrift fiir Metallkunde* **91**:4 (2000).

P. Chen, X. Wu, J. Lin, and K. L. Tan, “High H_2uptake by alkali-doped carbon nanotubes under ambient pressure and moderate temperatures,” *Science* **285**:91(1999).

RalphT.Yang,”Hydrogenstoragebyalkali-dopedcarbonnanotubes—revisited,” *Carbon* **38**:623 (2000). F. E. Pinkerton, B. G. Wicke, C. H. Olk, G. G. Tibbetts, G. P. Meisner, M. S. Meyer, and J. F. Herbst, “Thermogravimetric Measurement of hydrogen absorption in alkali-modified carbon materials,” *J. Phys. Chem. B* **104**:9460(2000).

V. Meregalli and M. Parrinello, “Review of theoretical calculations of hydrogen storage in carbon-based materials, *Appl. Phys. A* **72**:143-146(2001).

B. Kruse, S. Grinna and C. Buch, Bellona, “Hydrogen--Status and Possibilities”, June, 2002, 2002.http://www.bellona.no/en/energy/hydrogen/report_6-2002/index.html;http://www.bellona.no/en/energy/hydrogen/report_6-2002/22966.html3/22/03.

K. Taylor, Canadian Driver, “Special Report: The Future of Hydrogen” (2002). http://www.canadiandriver.com/articles/kt/hydrogen.htm3.23.03.

A. Bain and U. Schmidtchen, German Hydrogen Association, “Afterglow of aMyth. Why and How the ‘Hindenburg’ Burnt.””, January 18, 2000, 2000.http://www.dwv- info.de/pm/hindbg/hbe.htm#Gefahr3/23/03.

EV World, “Hydrogen Car Fire Surprise” (2003).http://www.evworld.com/databases/storybuilder.cfm?storyid=4823/22/03.

M. Swain, University of Miami, EV World, “Fuel Leak Simulation” (2003). http://evworld.com/library/Swainh2vgasVideo.pdf3/25/03.

“Hydrogen and the automobile,” Michael Chinworth, Automotive Engineering International, September, 2002,pp.89-95.

www.cartechdoe.gov/research/fuelcells/mgf-costs.html

www.cartechdoe.gov/research/fuelcells/cost-model.html

“FreedomCAR and Fuel Cells: Toward the Hydrogen Economy?” by Daniel Sperling, 9/27/02, unpublished. Also see his testimony, Transitioning from PNGV to FreedomCAR before the U.S. Congress, House Science Committee, Feb. 7,2002.

“Hydrogen Fuel May Be Clean; Getting It Here Looks Messy,” Jeffery Ball, The Wall Street Journal Online, March 7, 2003.http://online.wsj.com

BES Hydrogen Workshop, May 13-15, 2003.http://www.sc.doe.gov/bes/besac/BESAC_Dresselhaus_05-28-03.ppt

FURTIER READING

To assist members of POPA, and more generally, the members of the American Physical Society, a set of references and source material has been prepared.

"The Development of Ford's P2000 Fuel Cell Vehicle" Woong-Chul Yang, Ronald I. Sims, Bradford Bates, James A. Adams, SAE 2000-01-1061.

"Hydrogen storage in carbon nanostructures-still a long road from science to commerce?"

U. Bunger, W. Zittel, Applied Phys. A **72**, 147-151 (2001).

"Hydrogen separation and the direct high-temperature splitting of water" J. W. Warner and R. Stephen Berry, Int. J. Hydrogen Energy, Vol. 11, No. 2, pp. 91-100, 1996.

"Scenarios for the Future Energy Demand and CO_2-Emissions from the Global Transportation Sector" Stefan Hausberger, SAE 982216.

"Developments in fuel cells" Dan Carney, *Automotive Engineering International*, March 2002, pp. 47-52.

"Advance in fuel-cell development" Kevin Jost, Kami Bucholtz, Ray Gehm, *Automotive Engineering International*, June 2002, pp. 67-73.

"Engines for growth: Is this the beginning of the end for IC engines?" *Automotive Engineering International* June 2002, p. 25.

"Leading the way" Jack Yamaguchi, *Automotive Engineering International*, March 2003, p.54-58.

The September 2002 issue of *MRS Bulletin* is devoted to hydrogen storage. Guest editor is Louis Schlapbach.

Background Reading (Recommended by Ogden)

Accurex Environmental Corporation, 1996: *Evaluation of Fuel-Cycle Emissions on a Reactivity Basis*, 1, 2, prepared for the California Air Resources Board, ARB-Contract No. A166-134, Accurex Corporation, Mountain View, California, September 19.

Arthur D. Little, "Hydrogen Options for Fuel Cell Vehicles," presented at the 10^{th}Annual U.S. Hydrogen Meeting, Vienna, VA, April 3, 1999.

Bentley, J.M. et al. "The Impact of Electric Vehicles on CO2 Emissions," Arthur D. Little, Cambridge,MA, May 14, 1992.

Brogan, J. and S. R. Venkateswaran, " Diverse Choices for Electric and Hybrid Motor Vehicles: Implications for National Planners," presented at the Urban Electric Vehicle Conference, Stockholm, Sweden, may 25-27, 1992.

Darrow, K.G., "Light Duty Vehicle Fuel-Cycle Emission Analysis," prepared by Energy International, Inc., Bellevue, WA, for the Gas Research Institute, Chicago, IL, April 1994.

Darrow, K.G., "Comparison of Fuel-Cycle Emissions for Electric and Ultra-Low Emissions Natural Gas Vehicles," prepared by Energy International, Inc., Bellevue, WA, for the Southern California Gas Company, Los Angeles, CA, May 1994.

Delucchi, M.A., 1991: *Emissions of Greenhouse Gases from the Use of Transportation Fuels and Electricity*, Center for Transportation Research, Argonne National Laboratory, prepared for the United States Department of Energy Office of Transportation Technologies, Report No. ANL/ESD-TM-22, November.

Delucchi, M.A., 1993: *Emissions of Greenhouse Gases from the Use of Transportation Fuels and Electricity, **2** Appendices A-S,* Center for Transportation Research, Argonne National Laboratory, prepared for the United States Department of Energy Office of Transportation Technologies, Report No. ANL/ESD-TM-22, November.

Delucchi, M.A., 1997: *A Revised Model of Emissions of Greenhouse Gases from the Use of Transportation Fuels and Electricity,* Institute of Transportation Studies, University of California, Davis, CA, November.

General Motors Corporation, Argonne National Laboratory, BP Amoco, Exxon-Mobil, Shell, "Well-to-Wheel Energy Use and Greenhouse Gas Emissions of Advanced Fuel/Vehicle Systems: North American Analysis, Executive Summary Draft Report," April 2001.

National Renewable Energy Laboratory, "Fuel Cycle Evaluations of Biomass Ethanol and Reformulated Gasoline," October 1992.

J.M. Ogden, R.H. Williams, and E.D. Larson, "Exploring the Path Toward Near Zero Emissions in Transportation," draft report, October 2000.

P.L. Spath and M. K. Mann, " Lifecycle Assessment of Hydrogen Production via Natural Gas Steam Reforming," National Renewable Energy Laboratory Report NREL/TP-570- 27637, Golden, CO, November 2000.

Wang, M.Q. and H.-S. Huang, 1999: *A Full Fuel-Cycle Analysis of Energy and Emissions Impacts of Transportation Fuels Produced from Natural Gas,"* Center for Transportation Research, Argonne National Laboratory, prepared for the United States Department of Energy Office of Transportation Technologies, Report No. ANL/ESD-40. December.

M. A. Weiss, J. B. Heywood, E. D. Drake, A. Schafer and F.F. AuYeung, "On the Road in 2020," Massachusetts Institute of Technology Energy Laboratory Report No. MIT EL 00-003, October 2000.

Useful Internet Sites

Forum of Physics and Society subsite on Energy and Environment: http://www.aps.org/units/fps/energy.html

http://www.rff.org

http://www.pewclimate.org/projects/energy.cfm

http://www.stateline.org/

www.energy.gov/HQPress/releases02/janpr/FreedCarFactSheet_v.htm

http://www.eere.energy.gov/hydrogenandfuelcells/hydrogen/resources.html

http://www.epa.gov/newsroom/headline_021103.htm

The Hydrogen & Fuel Cell Investor at www.h2fc.com

The Hydrogen Web Page at www.hyweb.de
The Hydrogen and Fuel Cell Letter at www.hfcletter.com
The California Air Resources Board atwww.arb.ca.gov/homepage.htm
The California Fuel Cell Partnership at www.drivingthefuture.org
"Japan Auto Trends" is available at: http://www.jama.org

APPENDIX - A

From Ref.[8].

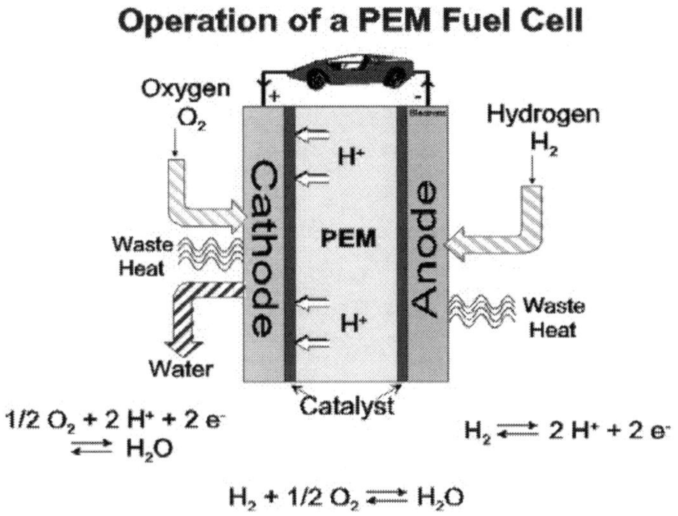

Principles of operation.

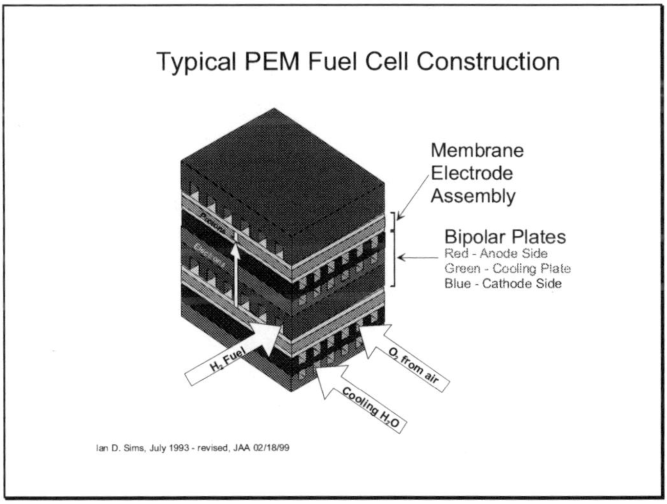

Schematic of a fuelcell.

APPENDIX - B

FreedomCAR vision and goals [2]:

Vision

Affordable full-function cars and trucks that are free of foreign oil and harmful emissions, without sacrificing safety, freedom of mobility, and freedom of vehicle choice.

Technology-Specific 2010 Goals[1]

1. To ensure reliable systems for future fuel cell powertrains with costs comparable with conventional internal combustion engine/automatic transmission systems,the goalsare
   - Electric propulsion system with a 15-year life capable of delivering at least 55 kW for 18 seconds and 30 kW continuous at a system costof $12/kW peak.
   - 60% peak energy-efficient, durable fuel cell power system (including hydrogen storage) that achieves a 325 W/kg power density and 220W/L operating on hydrogen. Cost targets are $45/kW by 2010, $30/kW by 2015.[2]
2. To enable clean, energy-efficient vehicles operating on clean, hydrocarbon-based fuels powered by either internal combustion powertrains or fuel cells, the goalsare
   - Internal combustion systems that cost $30/kW, have a peak brake engine efficiency of 45%, and meet or exceed emissions standards. [The termbrake engine efficiency refers to output shaftmeasurements.]
   - Fuel cell systems, including a fuel reformer, that have a peak brake engine efficiency of45% and meet or exceed emissions standards with a cost target of $45/kW by 2010 and $30/kW in 2015.[2,3]
3. To enable reliable hybrid electric vehicles that are durable and affordable, the goalis
   - Electric drivetrain energy storage with 15-year life at 300 Wh with discharge power of 25 kW for 18 seconds at a cost of$20/kW
4. To enable the transition to a hydrogen economy, ensure widespread availability of hydrogen fuels, and retain the functional characteristics of current vehicles, the goals are

- Demonstrated hydrogen refueling with developed commercial codes and standards and diverse renewable and non-renewable energy sources. Targets: 70% energy efficiency well-to-pump; cost of energy from hydrogen equivalent to gasoline at market price, assumed to be $1.25 per gallon (2001 dollars).[4]
- Hydrogen storage systems demonstrating an available capacity of 6 wt% hydrogen, specific energy of 2000W-h/kg, and energy density of 1100W-h/L at a cost of $5/kWh.[5]
- Internal combustion systems operating on hydrogen that meet cost targets of $45/kW by 2010 and $30/kW in 2015, have a peak brake engine efficiency of 45%, and meet or exceed emissions standards.

5. To improve the manufacturing base, the goalis
   - Material and manufacturing technologies for high-volume production vehicles that enable and support the simultaneous attainment of
     - 50% reduction in the weight of vehicle structure andsubsystems,
     - affordability, and
     - increased use of recyclable/renewable materials.

Notes:

1. Cost references are based on CY 2001 dollar values. Where power (kW) targets are specified, those targets are to ensure that technology challenges that would occur in a range of light-duty vehicle types would have to beaddressed.
2. Does not include vehicle tractionelectronics.
3. Includes fuel cell stack subsystem, fuel processor subsystem, and auxiliaries; does not include fueltank.
4. Targets are for hydrogen dispensed to a vehicle assuming a reforming, compressing, and dispensing system capable of dispensing 150 kg/day (assuming 60,000 SCF/day of natural gas is fed for reforming at the retail dispensing station) and servicing a fleet of 300 vehicles per day (assuming 0.5 kg used in each vehicle per day). Targets are also based on several thousand stations, and possibly demonstrated on several hundred stations. Technologies may also include chemical hydrides such as sodium borohydride.
5. Based on lower heating value of hydrogen; allows over a 300-milerange.

APPENDIX - C

From Ref.[49].

Reformer/Fuel Cell Cost Estimates				
	Annual Production Rate			
	500	10,000	30,000	500,000
Fuel Cell Stack	$17,258	$9,618	$9,394	$8,509
Air Loop	$1,160	$821	$734	$529
Water Loop	$1,106	$832	$757	$605
Coolant Loop	$620	$486	$450	$386
ATR	$3,531	$1,945	$1,532	$1,322
Reformate Loop	$1,172	$838	$739	$658
Fuel Loop	$879	$616	$573	$466
Controls	$719	$501	$442	$316
Misc./BOP	$320	$240	$220	$150
System Assembly	$723	$487	$442	$157
Total Cost	$27,489	$16,384	$15,282	$13,099
Cost/kW	$550	$328	$306	$262
Based on a 50kW-net system. All costs are preliminary, as DFMA optimization had not yet been completed. ATR and fuel cell stack examined in more detail than other system components. All costs include 10% cost contingency and markup to reflect profit, G&A.				

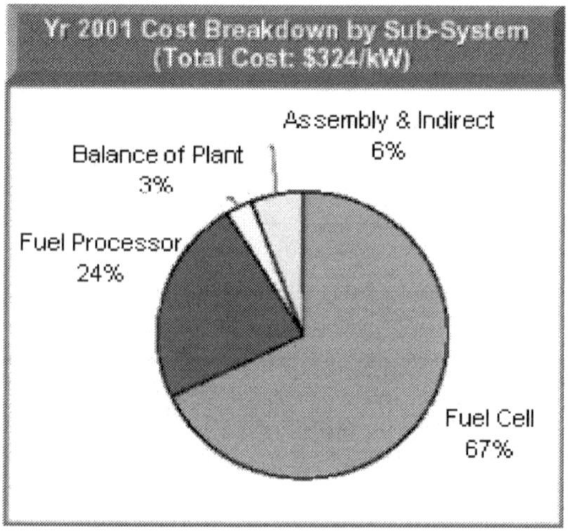

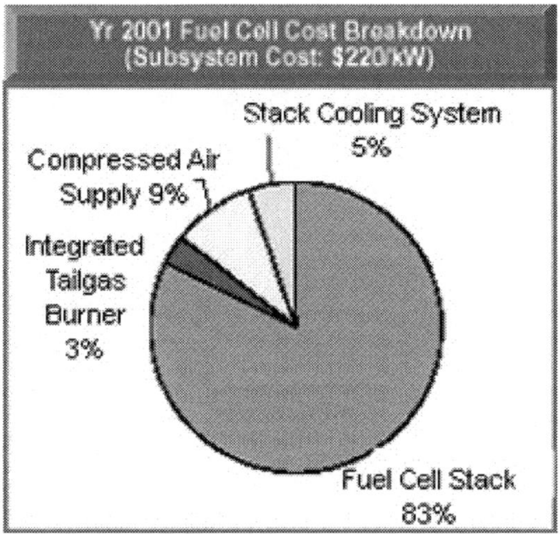

From Ref. [50].

Argonne National Lab Estimate for Central Production Facilities (quoted by Prof. Daniel Sperling, U. of California, Davis [51]):

- Infrastructure cost for producing H_2equivalent to 1.6 x 10^6barrels/day of oil (-20% of current automobile and light truckusage.)
- $400 billion forproduction
- $175 billion fordistribution
- [For comparison, the Exxon-Mobil market capitalization is $200 billion.]

BP Company Estimate for Outfitting Local Stations with Natural Gas Reformers

- $400,000 for reformer to make hydrogen using local gaslines
- $1.5 million to build conventional gas station From Ref.[52].

APPENDIX - D

Energy consumed in MJ/km (1 MJ/km = 1500 BTU/mi is equivalent to 0.013 gal/mi of gasoline.)

	2001 gasoline ICE	2020 gasoline ICE IEV	2020 diesel IEV	Fuel cell vehicle
Vehicle operation	2.47	1.07	0.92	0.54
Fuel cycle	0.52	0.22	0.13	0.42
Manufacturing	0.29	0.26	0.26	0.28
Total	3.28	1.55	1.31	1.24

[From Tables 8 and 9 of Ref. 21.]

The predicted well-to-wheels energy consumption for the 2020 fuel cell vehicle is 0.96 MJ/km = 1400 BTU/mi which is considerably less than the estimate in Table I. This reflects different assumptions about future technology development. For the conventional gasoline engine, the results also differ-4500 BTU/mi (3.0 MJ/km) compared to 6500 BTU/mi. The later calculation assumes a 22-mpg vehicle compared to 31 mpg here.

Life-Cycle Greenhouse Gas Emissions (gC/km).

	2001 gasoline ICE	2020 gasoline ICE IEV	2020 diesel IEV	Fuel cell vehicle
Vehicle operation	48.5	21.0	19.1	0
Fuel cycle	12.1	5.2	3.0	19.4
Manufacturing	5.5	5.0	5.1	5.3
Total	66.1	31.2	27.2	24.7

[From Tables 8 and 9 of Ref. 21.]

The emissions of greenhouse gas (GHG) in grams of carbon per kilometer traveled as given in Ref. 21 shows the significant improvements expected from advanced ICE HEVs and fuel cell vehicles. Life-cycle GHG emissions are approximately 20 g/MJ for the four vehicles tabulated in the two tables above.

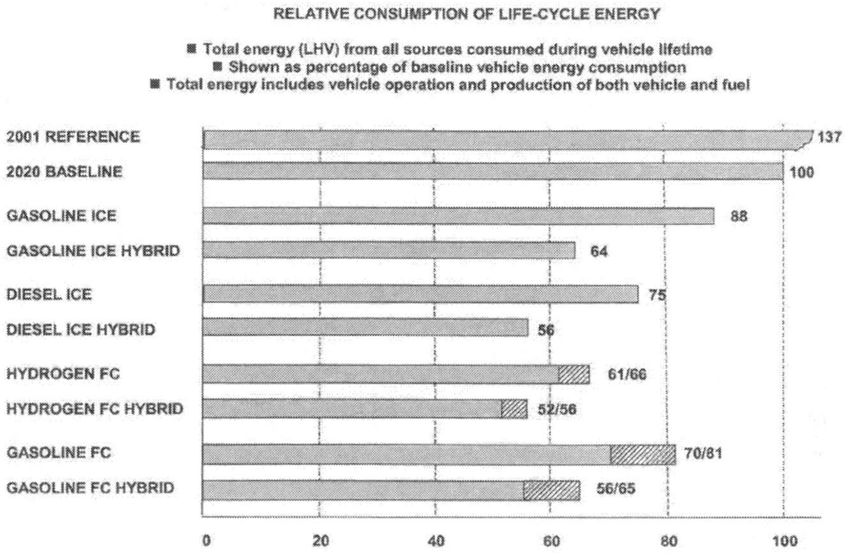
RELATIVE CONSUMPTION OF LIFE-CYCLE ENERGY
■ Total energy (LHV) from all sources consumed during vehicle lifetime
■ Shown as percentage of baseline vehicle energy consumption
■ Total energy includes vehicle operation and production of both vehicle and fuel
2001 REFERENCE
137
2020 BASELINE
100
GASOLINE ICE
88
GASOLINE ICE HYBRID
64
DIESEL ICE
75
DIESEL ICE HYBRID
56
HYDROGEN FC
61/66
HYDROGEN FC HYBRID
52/56
GASOLINE FC
70/81
GASOLINE FC HYBRID
56/65
0
20
40
60
80
100

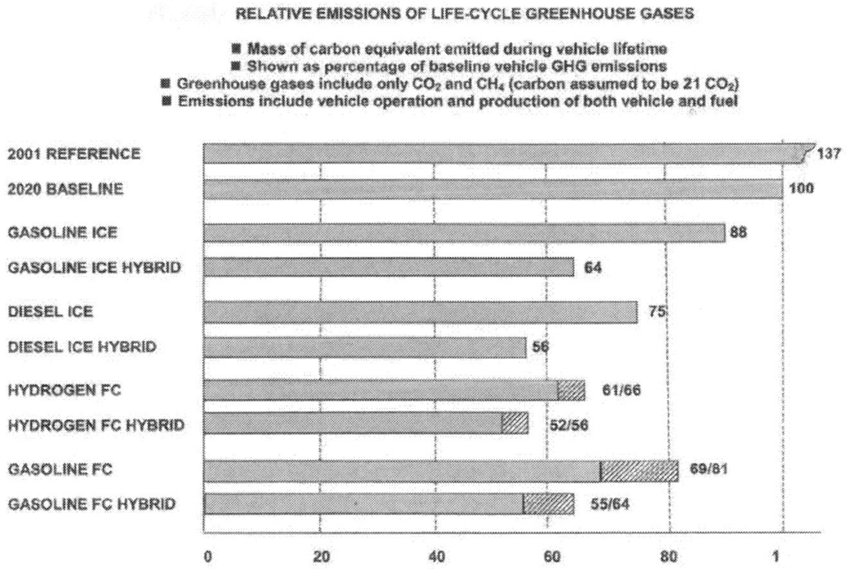
RELATIVE EMISSIONS OF LIFE-CYCLE GREENHOUSE GASES
■ Mass of carbon equivalent emitted during vehicle lifetime
■ Shown as percentage of baseline vehicle GHG emissions
■ Greenhouse gases include only CO_2 and CH_4 (carbon assumed to be 21 CO_2)
■ Emissions include vehicle operation and production of both vehicle and fuel
2001 REFERENCE
137
2020 BASELINE
100
GASOLINE ICE
88
GASOLINE ICE HYBRID
64
DIESEL ICE
75
DIESEL ICE HYBRID
56
HYDROGEN FC
61/66
HYDROGEN FC HYBRID
52/56
GASOLINE FC
69/81
GASOLINE FC HYBRID
55/64
0
20
40
60
80
1

From Ref. [21].

4

Hydrogen Consumption & Driving Range of Hydrogen Fuel Cell Vehicle under the CLTC-P Condition

Zhijie Duan

School of Environmental Science and Engineering, Tianjin University, Tianjin 300350, China;
Email: duanzhijie@gwm.cn

Harikrishna Nagwan

Quantum University, Dehradun Highway, Mandawar, Roorkee, Uttarakhand 247167, India

Dr. Seema Tiwari

Associate Professor Applied Science & General Engineering Army Institute of Technology, Pune.

Mr. John Abbas

IBIS Group, Lucknow, UP, India.

Prof. (Dr.) B. K. Sarkar (Patent Guru/ Expert)

MES, Geh Research, India, USA, Singapore, Japan, Austraila, Germany.

Prof. (Dr.) Reena Singh,

MES's, Pillai HOC College of Engineering &Technology, Rasayani Taluka Panvel,
Dist. Navi Mumbai, Maharashtra 410207, India

ABSTRACT

Hydrogen consumption and mileage are important economic indicators of fuel cell vehicles. Hydrogen consumption is the fundamental reason that restricts mileage. Since there are few quantitative studies on hydrogen consumption during actual vehicle operation, the high cost of hydrogen consumption in outdoor testing makes it impossible to guarantee the accuracy of the test. Therefore, this study puts forward a test method based on the hydrogen consumption of fuel cell vehicles under CLTC-P operating conditions to test the hydrogen consumption of fuel cell vehicles per 100 km.

Finally, the experiment shows that the mileage calculated by hydrogen consumption has a higher consistency with the actual mileage. Based on this hydrogen consumption test method, the hydrogen consumption can be accurately measured, and the test time and cost can be effectively reduced.

Keywords: *fuel cell; hydrogen consumption; driving range; test methods; comparative analysis*

INTRODUCTION

As global energy and environmental issues continue to emerge, hydrogen energy's zero-emission and pollution-free characteristics make it an ideal clean secondary energy source. Hydrogenfuelcellvehiclesarecurrently one of the most essential hydrogen energy application scenarios [1–3]. Environmental sustainability stimulates the development of electric vehicles with excellent energy-saving and emission reduction effects [4,5]. This has aroused great interest in new energy batteries. Major developed countries, such as the United States, Japan, South Korea, and Germany, as well as the European Union have vigorously developed hydrogen energy and fuel cell vehicle industries [6–9]. From the battery point of view, life evaluation and prediction help toextend the durability of fuel cells and accelerate their commercialization [10–12]. At the same time, from the perspective of the entire vehicle,the fuel economy of fuel cell vehicles cannot be ignored [13–15]. Therefore, the research on hydrogen consumption and the cruising range of fuel cell vehicles is also a hotissue.These two studies mainly focus on test method innovation, platform construction, and simulation modeling. The test methods of hydrogen consumption mainly include the temperature and pressure method, mass analysis method, and flow method [16]. Based on the calculation of fuel cell vehicles' actual hydrogen supply capacity, combined with the fundamental structure of the hydrogen supply system, Tian et al. [17] designed a test platform based on the weight method, temperature–pressure method, and flow method. Guo et al. [18] researched the fuel cell vehicle driving range test and evaluation method. The energy consumption and hydrogen consumption of electric-electric hybrid fuel cell vehicles are tested, forming a driving range evaluation method for different fuel cell vehicles. Considering the test and analysis results, suggestions and references for improving the driving range of fuel cell vehicles are put forward. J.S. et al. [19] proposed a method for measuring the fuel consumption of hydrogen fuel cell vehicles on a chassis dynamometer, which provides information for the development of hydrogen fuel economy test methods.

In terms of simulation modeling, the main research point is to improve the accuracy of the hydrogen consumption model. Hou [20] established a semi-empirical kinetic model of hydrogen consumption in a fuel cell engine and conducted experimental verification based on the steady-state hydrogen consumption model. The results show that the model has high accuracy, a simple structure, and few parameters and can better reflect the dynamic characteristics of hydrogen consumption of fuel cell engines. Javier et al. [21] calculated the hydrogen consumption of hybrid electric vehicles and simulated FC hybrid electric vehicles under standard operating conditions, developing an all-electric model and a thermal model of the entire vehicle. Fang et al. [13] designed a hydrogen consumption measurement research platform for a proton exchange membrane fuel cell vehicle (PEMFCV). They used the system to conduct multiple scheduled tests to verify that the platform meets the research needs. Hou et al. [22] introduced a dynamic model of hydrogen consumption in a fuel cell stack. Under the step load change, the hydrogen consumption changes with a specific time lag. The delay time of the load-up phase is shorter than that of the load-down stage. The overshoot duration and peak value are distributed in a normal range. Other studies to carry out hydrogen consumption prediction of the fuel cell system based on the hybrid intelligent method have achieved satisfactory results in the verification process of the model [23–25].

However, the above research is limited to studying the hydrogen consumption model and consumption alone, without linking hydrogen consumption with a driving range and lacking consistency between test and calculated data. This article takes Test vehicle A and Test vehicle B as the research objects to conduct hydrogen consumption and driving range tests of fuel cell vehicles to study the consistency of vehicle hydrogen consumption and moving range test data. This work can provide a data reference for supplementing or revising the hydrogen fuel cell vehicle standards. More importantly, referring to the above method to carry out the hydrogen consumption/driving range test for fuel cell vehicles, the testing time and cost will be significantly reduced on the premise of ensuring the testing accuracy.

RESEARCH OBJECT

The research object of this article is based on two models, test vehicles A and B. The Type A fuel cell vehicle has a maximum fuel cell power of 114 kW and a battery power energy of 1.65 kWh/5.94 MJ. Type B fuel cell vehicles have a fuel cell system power of 95 kW and a power battery of 1.56 kWh/5.62 MJ. Two kinds of hydrogen storage-related parameters are listed in Table1.

Table 1. Hydrogen storage parameters of car models.

Test Vehicle Model	Hydrogen Storage Pressure	Hydrogen Storage Volume	Hydrogen Storage	Driving Range (NEDC)	Fuel Cell Power (kW)
Test vehicle A	70 Mpa	122.4 L	5.0 kg/599.8 MJ	650 km	114
Test vehicle B	70 Mpa	156.6 L	6.33 kg/759.3 MJ	800 km	95

TEST METHOD

The test environment meets the requirements of the GB/T39132-2020 fuel cell electric vehicle type test procedure of the driving range test method. The test room is equipped with air-conditioning, and the test environment temperature is 20~30 C. The two test vehicles were running-in. The running-in mileage exceeded 1000 km, and 300 km within seven days before the test. The test vehicle A chassis dynamometer's load setting follows the vehicle seven days before the test. The test vehicle A chassis dynamometer's load setting follows the vehicle reference mass specified in GB18352.5–2013, and test vehicle B chassis dynamometer load is set under the provisions in GB18352.5–2016. The test parameters are listed in Table 2, and the test flowchartis shown in Figure 1.

Table 2. Specific parameters of test conditions.

Model		Loading Convention	Coefficient A	Coefficient B	Coefficient C
Test vehicle A	DYNO	1930 kg	8.50 N	0 N/(km/h)	0.05770 N/(km/h)^2
Test vehicle B	ROAD	2030.3 kg	156.85 N	1.3640 N/(km/h)	0.03630 N/(km/h)^2
	DYNO	2030.3 kg	−30.46 N	1.7424 N/(km/h)	0.03290 N/(km/h)^2

Figure 1: The test flowchart.

Introduction to the Test Methods for Hydrogen Consumption of Fuel Cell Vehicles

The New European Driving Cycle (NEDC) is a standard car cycle defined based on the road conditions in the European Union. Before 2020, most of the test cycle conditions of the country's automobiles adopt NEDC conditions. It is very different from the actual road conditions. The tested fuel consumption and emission data are also different from the existing conditions. Therefore, the application has significant limitations. The NEDC driving cycle curve is shown in Figure 2.

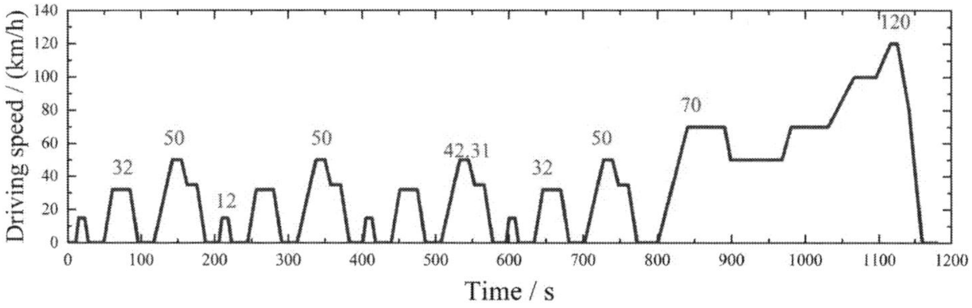

Figure 2: NEDC working condition curve.

Both Test vehicles A and B belong to category A vehicles. CLTC-P is a Chinese light-duty vehicle test cycle passenger car. First, the vehicle is preheated (1 CLTC-P cycle workingcondition,standingfor15min),andth en6CLTC-Pcycleworkingconditionsare established to complete the test. CLTC-P reflects the actual road conditions with Chinese characteristics more truly, including a more reasonable definition of average speed and maximum speed, more comprehensive driving conditions, a better proportion ofparking modes, and richer dynamic acceleration and deceleration conditions. Figure 3 shows the driving curve of CLTC-P Chinese passengercars.

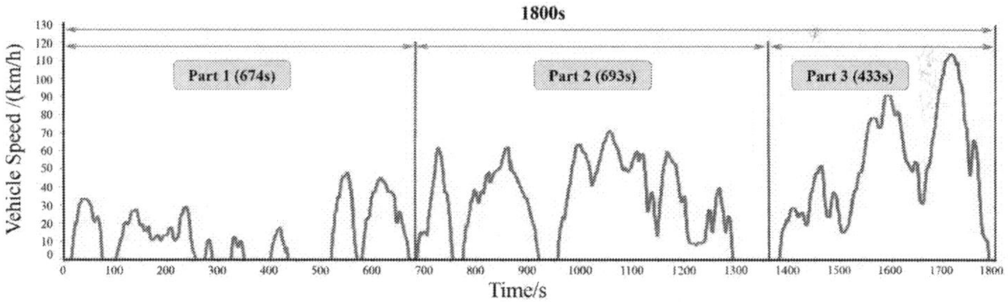

Figure 3: CLTC-P working condition curve.

Introduction of Fuel Cell Vehicle Driving Range Test Method

Fuel cell electric vehicle tests are carried out in accordance with the CLTC-P cycle conditions specified in GB/T38146.1. The car will stop the test if it meets one of the following two conditions:

- The vehicle meter will give a parkingindication;
- The vehicle speed tolerance does not meet ±2 km/h or the time tolerance exceeds 10s.

When the vehicle reaches the test end condition, keep the vehicle gear unchanged. Then, wait for the vehicle to coast to the lowest stable speed or 5 km/h, and depress the brake pedal to stop. After the test is over, the vehicle's mileage is rounded to an integer value to be the driving mileage of the test.

Test Method and Procedure

The test device and interface diagram are shown in Figure 4 [26]. The NEDC cycle conditions were not selected for the hydrogen consumption test for data comparison. The hydrogen consumption is consistent with the driving range test, and both use CLTC-P cycle conditions. To fully accumulate test samples and observe more comprehensive test results and comparisons, hydrogen consumption has been continuously carried out in 13 CLTC-P cycle conditions. Simultaneously, Test vehicle B cannot be connected to the test room to supply hydrogen, so the hydrogen consumption test uses vehicle-mounted hydrogen. The data were obtained according to the temperature and pressure method. The pressure and temperature data were collected by the sensor of the vehicle-mounted hydrogen storage bottle through the vehicle CAN communication.

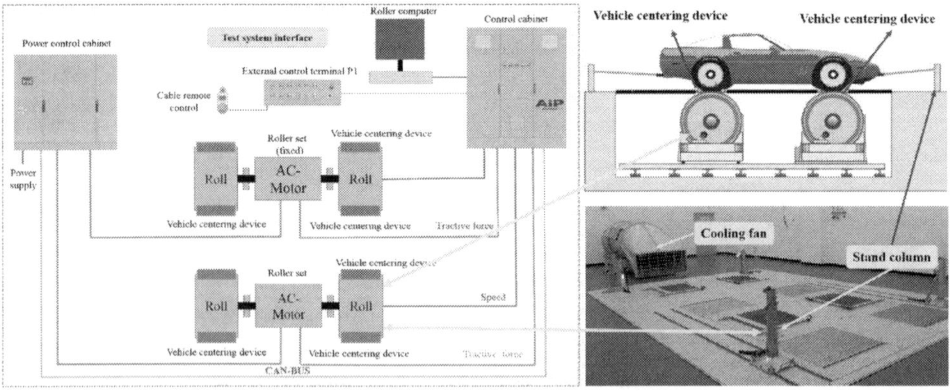

Figure 4: The diagram of the test device andinterface.

Hydrogen Consumption Test for Fuel Cell Vehicles

Both models have undergone 13 CLTC-P cycle test conditions. The specific test steps are asfollows:

1. Test vehicle A vehicles close the on-board hydrogen supply system and use the hydrogen from the hydrogen supply pipeline outside the test room. In this study, Austrian AVL's hydrogen flow metering system was used to measure the value of vehicle

hydrogen consumption. Test vehicle B vehicles use the pressure-temperature method of the vehicle's hydrogen storage tank to measure hydrogen consumption indirectly;

2. After the vehicle is fixed to the chassis dynamometer, leave it at room temperature (20–30 ◦C) for 30min;
3. Vehicle road load and chassis dynamometer load are set following national standards;
4. The vehicle is warmed up, the vehicle is driven on the chassis dynamometer, the vehicle is turned off after the CLTC-P cycle is completed, and the vehicle is set to the initial value for 15 min;
5. Start the vehicle signal acquisition equipment. The acquired signals include but are not limited to each hydrogen cylinder temperature, hydrogen cylinder pressure, hydrogen fuel cell stack current, voltage, vehicle mileage, hydrogen consumption, and hydrogen flow, and the acquisition frequency of each signal is 10 Hz;
6. Start the vehicle and perform 13 continuous working condition tests in accordance with the CLTC-P working condition table. The vehicle cannot be stopped during the test operation until the end of the test;
7. Record the mileage at the end of each test cycle and keep the mileage to one decimal place. If the hydrogen flow rate is used to test the hydrogen consumption, the hydrogen consumption is recorded at the same time, and the hydrogen consumption is kept to one decimal place;
8. After the vehicle completes 13 cycles of testing, record the total mileage and stop the vehicle;
9. Stop data recording and analyze the data.

Hydrogen Consumption Test for Fuel Cell Vehicles

1. Fill the vehicle-mounted hydrogen cylinder with hydrogen until the rated pressure is70 MPa;
2. Keep the car static at room temperature (20–30 C) for 30 min, then fix the car to the chassis dynamometer;
3. Start the vehicle, test according to CLTC-P working conditions, and stop when the test end conditions are reached;
4. Record the mileage and accumulated mileage of each cycle working condition, keeping one decimal place for the mileage of each working condition. The accumulated mileage is rounded up.

TEST RESULTS AND DISCUSSION

Analysis and Discussion of Hydrogen Consumption Test Data

The vehicle test continuously carried out 13 CLTC-P cycle conditions and obtained hydrogen consumption per 100 km for 11 groups of three continuous CLTC-P cycle conditions and eight groups of hydrogen consumption per 100 km for six consecutive CLTC-P cycles. The actual hydrogen consumption per 100 km is calculated based on the actual hydrogen consumption of each cycle and the corresponding mileage. The relative deviation is calculated based on the actual value and the average value of hydrogen consumption ((Actual value – average value)/ Average value 100%).

4.1.1. Analysis of Test Vehicle A Test Results and Hydrogen Consumption Measurement Method

The actual hydrogen consumption of each CLTC-P cycle of Test vehicle A is measured using hydrogen from the hydrogen supply pipeline outside the test room and measured by the hydrogen flow metering system of AVL, Austria. The specific data are shown in Table 3.

According to the data in the above table, the hydrogen consumption per hundred kilometers of a single cycle is 0.977 kg/100 km, the hydrogen consumption of three cycles is 0.979 kg/100 km, and the hydrogen consumption of six cycles is 0.983 kg/100 km. The maximum deviation of hydrogen consumption in a single working condition from the average value is 3.59%. The maximum deviation of the hydrogen consumption from the average value for three consecutive cycles is 3.0%. The maximum deviation of hydrogen consumption from the average value for six consecutive cycles is 1.0%. The deviation distribution of each cycle is shown in Figure 5.

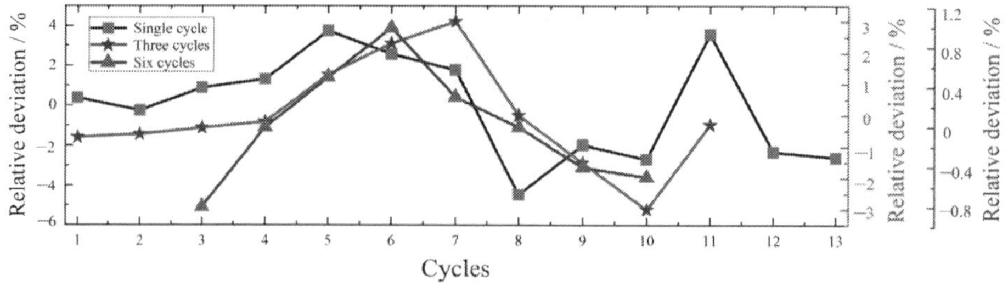

Figure 5: Cyclic deviation distribution.

Table 3: Actual hydrogen consumption of Test vehicle A under the CLTC-P cycle.

Cycles	Cycle Mileage (km)	Actual Hydrogen Consumption (g)	Actual Hydrogen Consumption per Hundred Kilometers (kg/100 km)	Relative Difference Rate %
1	14.6	141.7	0.9702	0.38
2	14.4	140.8	0.9775	−0.25
3	14.5	142.4	0.9819	0.89
4	14.5	143.0	0.9860	1.32
5	14.4	146.4	1.0167	3.75
6	14.6	144.7	0.9911	2.57
7	14.3	143.6	1.0042	1.78
8	14.4	134.8	0.9361	−4.46
9	14.4	138.3	0.9604	−1.97
10	14.4	137.3	0.9535	−2.69
11	14.4	146.2	1.0146	3.59
12	14.4	137.8	0.9569	−2.32
13	14.4	137.5	0.9545	−2.60
Average value	14.4	141.1	0.9772	——

Cycles	Cycle Mileage (km)	Actual Hydrogen Consumption (g)	Actual Hydrogen Consumption per Hundred Kilometers (kg/100 km)	Relative Difference Rate %
1~3	43.5	423.1	0.973	−0.7
2~4	43.4	422.5	0.974	−0.6
3~5	43.4	423.1	0.975	−0.4
4~6	43.5	424.9	0.977	−0.2
5~7	43.3	429.6	0.992	1.3
6~8	43.3	433.5	1.001	2.3
7~9	43.1	434.7	1.009	3.0
8~10	43.2	423.1	0.979	0.0
9~11	43.2	416.7	0.965	−1.5
10~12	43.2	410.4	0.950	−3.0
11~13	43.2	421.7	0.976	−0.3
Average value	43.3	423.9	0.9791	——

Cycles	Cycle Mileage (km)	Actual Hydrogen Consumption (g)	Actual Hydrogen Consumption per Hundred Kilometers (kg/100 km)	Relative Difference Rate %
1~6	87.0	848.0	0.9747	−0.8
2~7	86.7	852.1	0.9828	0.0
3~8	86.7	856.6	0.9880	0.5
4~9	86.6	859.6	0.9926	1.0
5~10	86.5	852.7	0.9858	0.3
6~11	86.5	850.2	0.9829	0.0
7~12	86.3	845.1	0.9793	−0.4
8~13	86.4	844.8	0.9778	−0.5
Average value	86.6	851.1	0.9830	——

Analysis of Test Vehicle B Test Results and Hydrogen Consumption Evaluation Method

The actual hydrogen consumption of each CLTC-P cycle of Test vehicle Buses vehicle- mounted hydrogen. It is calculated and measured following the pressure-temperature method with the vehicle hydrogen storage bottle. The vehicle hydrogen storage bottle's pressure and temperature sensor signals are collected and output by the vehicle CAN signal. Limited by the accuracy of the pressure and temperature sensors of the vehicle's hydrogen storage cylinder, the hydrogen consumption varies significantly among various cycle conditions. The specific data are shown in Table 4.

Table 4: Actual hydrogen consumption of Test vehicle B under CLTC-P cycle.

Cycles	Cycle Mileage (km)	Actual Hydrogen Consumption (g)	Actual Hydrogen Consumption per Hundred Kilometers (kg/100 km)	Relative Difference Rate %
1	14.5	118.8	0.8193	−1.0
2	14.4	135.7	0.9424	13.9
3	14.3	137.3	0.9601	16.0
4	14.3	91.6	0.6406	−22.6
5	14.4	122.9	0.8535	3.1
6	14.3	115.3	0.8063	−2.6
7	14.5	121.1	0.8352	0.9
8	14.4	118.0	0.8194	−1.0
9	14.4	95.8	0.6653	−19.6
10	14.2	124.8	0.8796	6.3
11	14.5	113.0	0.7793	−5.8
12	14.5	135.7	0.9359	13.1
13	14.2	116.8	0.8225	−0.6
Average value	14.2	119.0	0.8277	——

Cycles	Cycle Mileage (km)	Actual Hydrogen Consumption (g)	Actual Hydrogen Consumption per Hundred Kilometers (kg/100 km)	Relative Difference Rate %
1~3	43.2	391.8	0.907	10.4
2~4	43.0	364.6	0.848	3.2
3~5	43.0	351.8	0.818	−0.4
4~6	43.0	329.8	0.767	−6.7
5~7	43.2	359.3	0.832	1.2
6~8	43.2	354.4	0.820	−0.2
7~9	43.3	334.9	0.773	−5.9
8~10	43.0	338.7	0.788	−4.1
9~11	43.1	333.7	0.774	−5.8
10~12	43.2	373.6	0.865	5.2
11~13	43.2	365.5	0.846	3.0
Average value	43.1	354.4	0.8217	——

Cycles	Cycle Mileage (km)	Actual Hydrogen Consumption (g)	Actual Hydrogen Consumption per Hundred Kilometers (kg/100 km)	Relative Difference Rate %
1~6	86.2	721.6	0.8371	2.9
2~7	86.2	723.9	0.8398	3.2
3~8	86.2	706.2	0.8193	0.7
4~9	86.3	664.7	0.7702	−5.3
5~10	86.2	698.0	0.8097	−0.5
6~11	86.3	688.1	0.7973	−2.0
7~12	86.5	708.5	0.8191	0.7
8~13	86.2	704.2	0.8169	0.4
Average value	86.3	701.9	0.8137	—

According to the data in the above table, the hydrogen consumption per hundred kilometers of a single cycle is 0.828 kg/100 km, the hydrogen consumption of three cycles is 0.822 kg/100 km, and the hydrogen consumption of six cycles is 0.815 kg/100 km. The maximum deviation of hydrogen consumption in a single working condition from the average value is −22.6%. The maximum deviation of hydrogen consumption from the average value for three consecutive cycles is 10.4%. The maximum deviation of hydrogen consumption from the average value for six consecutive cycles is 5.3%. The relative deviation decreases as the number of cycle conditions increases. The deviation distribution of each process is shown in Figure 6.

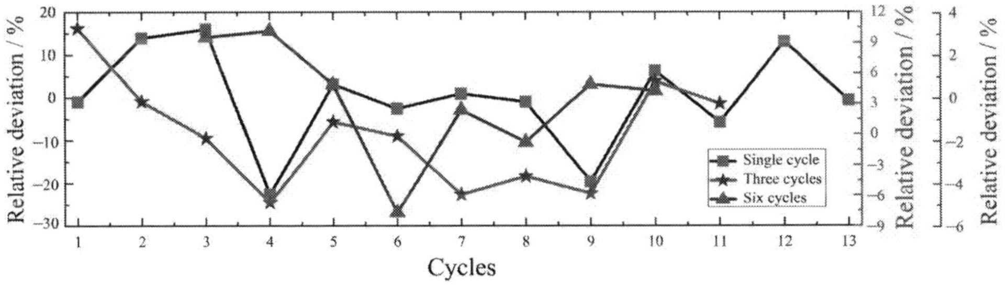

Figure 6: Cyclic deviation distribution.

Analysis and Discussion of Test Data of Vehicle Driving Range

The vehicle's driving range is based on the current national standard fuel cell electric vehicle test method. CLTC-P cycle conditions were used to obtain the actual test. At the same time, according to the test, the hydrogen consumption of the vehicle per 100 km (calculated by the trial and test of six consecutive CLTC-P cycle conditions) and the vehicle's available hydrogen amount is indirectly calculated to obtain the driving range (computed value). Hence, 13 successive CLTC-P cycle conditions are received in the actual test vehicle. Eight groups of six continuous

CLTC-P cycle conditions are obtained for hydrogen consumption per 100 km. The driving range (calculated value) is obtained.

Analysisof Test Vehicle A Test Results and Driving Range Test Method

The actual hydrogen consumption of each CLTC-P cycle of Test vehicle A uses the hydrogen of the hydrogen supply pipeline outside the test room. It is measured by the hydrogen flow metering system of Austria AVL Company, Melbourne, Australia. The specific values are shown in Table 5.

Table 5: Driving range of Test vehicle A under CLTC-P cycle conditions.

Serial Number	Cycle Condition	Hydrogen Consumption (g)	Vehicle Mileage (km)	Hydrogen Consumption per Hundred Kilometers (kg/100 km)	Driving Range (Calculated Value) (km)	Difference Rate %
1	1~6	848.0	87.0	0.975	513	0.6
2	2~7	852.1	86.7	0.983	509	−0.2
3	3~8	856.6	86.7	0.988	506	−0.8
4	4~9	859.6	86.6	0.993	504	−1.2
5	5~10	852.7	86.5	0.986	507	−0.5
6	6~11	850.2	86.5	0.983	509	−0.3
7	7~12	845.1	86.3	0.979	511	0.1
8	8~13	844.8	86.4	0.978	511	0.3

Test driving range (km): 510 km.

According to the data in the above table, the hydrogen consumption per hundred kilometers for six cycles is 0.978kg/ 100km.The maximum deviation between the continuous driving range and the average value of six stable cycle conditions is −1.2%, and the deviation distribution of each cycle is shown in Figure 7.

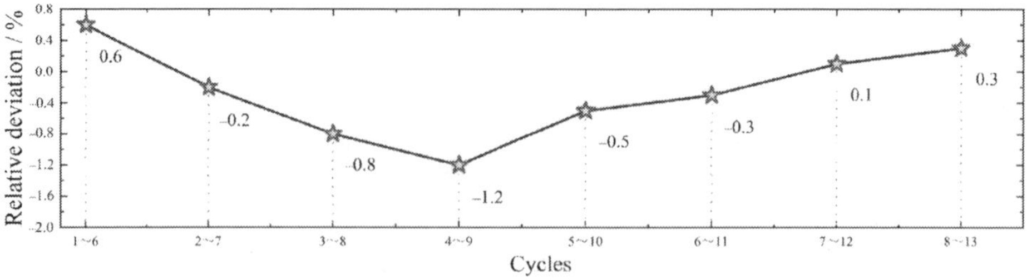

Figure 7: Deviation distribution of 6 continuous cycle conditions.

Analysis of Test Vehicle B Test Results and Driving Range Test Method

The actual hydrogen consumption of the Test vehicle B vehicle CLTC-P cycle working condition uses the vehicle-mounted hydrogen, which is calculated and measured by the pressure-temperature method of the vehicle hydrogen storage bottle. The vehicle hydrogen storage bottle's pressure and temperature sensor signals are collected and output by the vehicle CAN signal. The specific values are shown in T

Table 6: Driving range of Test vehicle B under CLTC-P cycle conditions.

Serial Number	Cycle Condition	Hydrogen Consumption (g)	Vehicle Mileage (km)	Hydrogen Consumption per Hundred Kilometers (kg/100 km)	Driving Range (Calculated Value) (km)	Difference Rate %
1	1~6	721.6	86.2	0.837	756	−5.7
2	2~7	723.9	86.2	0.840	754	−6.0
3	3~8	706.2	86.2	0.819	773	−3.7
4	4~9	664.7	86.3	0.770	822	2.5
5	5~10	698.0	86.2	0.810	782	−2.5
6	6~11	688.1	86.3	0.797	794	−1.0
7	7~12	708.5	86.5	0.819	773	−3.6
8	8~13	704.2	86.2	0.817	775	−3.4

Test driving range (km): 802 km.

According to the driving range (calculated value), hydrogen per hundred kilometers for six cycles is 0.817 kg/100 km. Given the test driving range and difference rate ((continuing driving range (calculated value) − test driving range)/test driving range × 100%), the maximum difference rate of each group of values is − 6.0%. The specific distribution is shown in Figure 8..

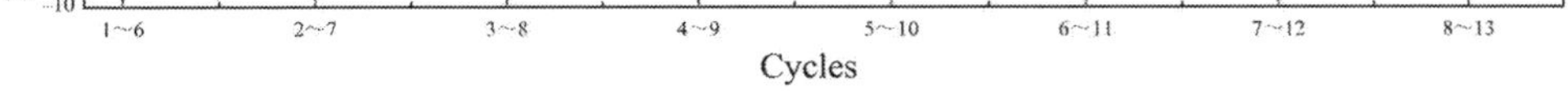

Figure 8: Deviation distribution of 6 continuous cycle conditions.

Energy-Saving Effect of the Braking Energy Recovery Strategy

This section compares the braking energy recovery strategy and the non-brake energy recovery strategy and the non-brake energy recovery strategy adopted by the Test vehicle B vehicle under the CLTC-P cycle. From Table 7 and Figure 9, it can be seen that the initial driving range increased by 135.8 km, and the driving range at termination was increased by 175.5 km after adopting the braking energy recovery strategy. The SOC values of the batteries with and without the braking energy recovery strategy are 57% and 51%, respectively, an increase of 6%.

Table 7: Parameter data of Test vehicle B's two strategies under CLTC-P conditions.

Main Parameters	With Braking Energy Recovery	Without Braking Energy Recovery
Initial driving range	612.4 km	476.6 km
Final driving range	463.3 km	287.8 km
Battery SOC	57%	51%

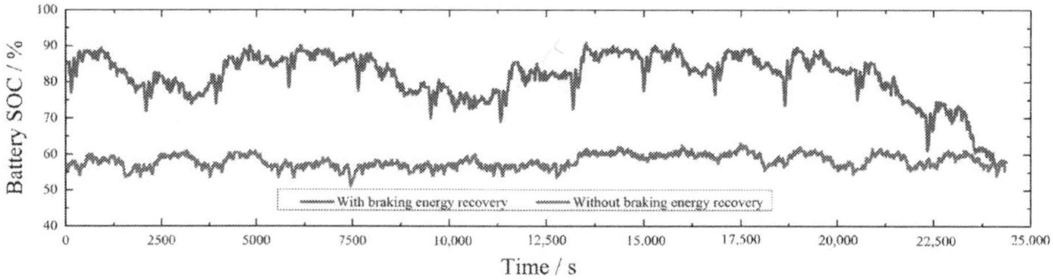

Figure 9: Comparison of battery SOC value changes.

CONCLUSIONS

Six consecutive standard cycles of external hydrogen supply were used to test the hydrogen consumption of fuel cell vehicles during CLTC-P operation. The difference between the calculated and tested driving ranges was less than 1.2%. The cycle conditions are divided into three groups, and the mileage is calculated by calculating the hydrogen consumption per hundred kilometers of each group. The difference between the actual mileage and the calculated average mileage decreases as the number of cycles increases.

After adopting the braking energy recovery strategy, the initial driving range was increased by 135.8 km, and the driving range at the end was increased by 175.5 km. The SOC values of the batteries with and without the braking energy recovery strategy are 57% and 51%, respectively, an increase of 6%.

Referring to the above method to carry out the hydrogen consumption/driving range test for fuel cell vehicles and calculate the driving range value of fuel cell vehicles, the fuel cell vehicle driving range test time and the test cost are significantly reduced. At the same time, the testing unit and the country of the testing organization can also guarantee the accuracy of the test values.

REFERENCES

Mathieu, T.; Kourosh, M.; Asmae, M. Techno-Economics of a New High Throughput Process for Proton Exchange Membranes Manufacturing. *World Electr.* Veh. J. 2016, 8, 431–442.

Speers, P. Hydrogen Mobility Europe (H2ME): Vehicle and Hydrogen Refuelling Station Deployment Results. *World Electr.* Veh. J. 2018, 9, 2. [CrossRef]

Zhou, S.; Jin, J.; Wei, Y. A Driving Cycle for a Fuel Cell Logistics Vehicle on a Fixed Route: Case of the Guangdong Province. *World Electr.* Veh. J. 2021, 12, 5. [CrossRef]

Hong, J.; Wang, Z.; Chen, W.; Wang, L.; Lin, P.; Qu, C. Online Accurate State of Health Estimation for Battery Systems on Real-World Electric Vehicles with Variable Driving Conditions Considered. J. Clean. Prod. 2021, 294, 125814. [CrossRef]

Hong, J.;Wang, Z.; Yao, Y. Fault prognosis of battery system based on accurate voltage abnormity prognosis using long short-term memory neural networks. *Appl. Energy* 2019, 251, 113381. [CrossRef]

Han, J.; Dongsuk, K.; Youngjin, P. Sensitivity analysis for assessing robustness of position- based predictive energy management strategy for fuel cell hybrid electric vehicle. *World Electr.* Veh. J. 2015, 7, 330–341. [CrossRef]

Andaloro, L.; Micari, S.; Napoli, G.; Polimeni, A.; Antonucci, V. A Hybrid Electric Fuel Cell Minibus: Drive Test. *World Electr.* Veh. J. 2016, 8, 131–138. [CrossRef]

Kim, N.; Moawad, A.; Vijayagopal, R.; Rousseau, A. Impact of Fuel Cell and Storage System Improvement on Fuel Consumption and Cost. *World Electr.* Veh. J. 2016, 8, 305–314. [CrossRef]

Cullen, D.A.; Neyerlin, K.C.; Ahluwalia, R.K.; Mukundan, R.; More, K.L.; Borup, R.L.; Weber, A.Z.; Myers, D.J.; Kusoglu, A. New roads and challenges for fuel cells in heavy-duty transportation. *Nat. Energy* 2021, 6, 462–474. [CrossRef]

Pei, P.; Chen, D.;Wu, Z.; Ren, P. Nonlinear methods for evaluating and online predicting the lifetime of fuel cells. *Appl. Energy* 2019, 254, 113730. [CrossRef]

Szałek, A.; Pielecha, I.; Cieslik, W. Fuel Cell Electric Vehicle (FCEV) Energy Flow Analysis in Real Driving Conditions (RDC). *Energies* 2021, 14, 5018. [CrossRef]

Pielecha, I.; Cie´slik, W.; Szałek, A. The use of electric drive in urban driving conditions using a hydrogen powered vehicle-Toyota Mirai. Combust. Engines 2018, 57, 51–58. [CrossRef]

Maodong, F.; Mingjie, C.; Qingchun, L.; Zhenhua, J. Hydrogen Consumption Measurement Research Platform for Fuel Cell Vehicles. In Proceedings of the International Conference on Electrical and Control Engineering, Wuhan, China, 25–27 June 2010; IEEE: Piscataway, NJ, USA, 2010; pp. 1142–1145.

Ahluwalia, R.K.; Wang, X.; Rousseau, A.; Kumar, R. Fuel economy of hydrogen fuel cell vehicles. J. *Power Sources* 2004, 130, 192–201. [CrossRef]

Offer, G.J.; Contestabile, M.; Howey, D.A.; Clague, R.; Brandon, N.P. Techno-economic and behavioural analysis of battery electric, hydrogen fuel cell and hybrid vehicles in a future sustainable road transport system in the UK. *Energy Policy* 2011, 39, 1939–1950. [CrossRef]

Lai, X.; Ling, L.; Liang, W.; Wang, B. Test Platform Development for Fuel Cell Vehicle's Hydrogen Management System. TELKOMNIKA Indones. J. *Electr. Eng.* 2012, 10. [CrossRef]

Tian, Y.; Zhang, Y.F.; Jin, Z.H.; Wang, K.L.; Nie, S.F.; Lu, Q.C. Development of Hydrogen Consumption Test Platform for Fuel Cell Vehicles. *Adv. Mater. Res.* 2013, 602–604, 1031–1035. [CrossRef]

Guo, T. Research on the Measurement Method of Fuel Cell Vehicle Driving Range. Power Technol. 2021, 45, 848–850+931. 19. Lim, J.-S.; Lee, H.-W.; Hong, Y.-S.; Lee, K.-B.; Yong, G.-J.; Kwon, H.-B. Development on Fuel Economy Test Method for Hydrogen Fuel Cell Vehicles. Trans. Korean Hydrog. *New Energy Soc.* 2010, 21, 207–213.

Hou, Y.; Zhuang, M.; Wan, G. A study on a semi-empirical dynamic model for hydrogen consumption of a fuel cell engine. *Automot. Eng.* 2007, 29, 942–945.

Pino, F.J.; Marcos, D.; Bordons, C.; Rosa, F. Car air-conditioning considerations on hydrogen consumption in fuel cell and driving limitations. *Int. J. Hydrog.* 2015, 40, 11696–11703. [CrossRef]

Hou, Y.; Shen, C.; Hao, D.; Liu, Y.; Wang, H. A dynamic model for hydrogen consumption of fuel cell stacks considering the effects of hydrogen purge operation. *Renew. Energy* 2014, 62, 672–678. [CrossRef]

Montero-Sousa, J.A.; Aláiz-Moretón, H.; Quintián, H.; González-Ayuso, T.; Novais, P.; Calvo-Rolle, J.L. Hydrogen consumption prediction of a fuel cell based system with a hybrid intelligent approach. Energy 2020, 205, 117986. [CrossRef] *World Electr.* Veh. J. 2022, 13, 9 13 of 13

Li, Y.; Pei, P.;Wu, Z.; Ren, P.; Jia, X.; Chen, D.; Huang, S. Approaches to avoid flooding in association with pressure drop in proton exchange membrane fuel cells. *Appl. Energy* 2018, 224, 42–51. [CrossRef]

Chen, D.; Pan, L.; Pei, P.; Huang, S.; Ren, P.; Song, X. Carbon-coated oxygen vacancies-rich Co3O4 nanoarrays grow on nickel foam as efficient bifunctional electrocatalysts for rechargeable zinc-air batteries. *Energy* 2021, 224, 120142. [CrossRef]

Duan, Z.; Zhang, L.; Feng, L.; Yu, S.; Jiang, Z.; Xu, X.; Hong, J. Research on Economic and Operating Characteristics of Fuel Cell Cars Based on Real Vehicle Tests. *Energies* 2021, 14, 7856. [CrossRef]

5

Research on Hydrogen Energy and Fuel Cell Vehicle Roadmap in Various Countries

Dr. Nimish H. Vasoya

Department of Balbhavan, Children's University Gandhinagar, Email: hod.bal@cugujarat.ac.in

Jiang Yunzhe

1Strategic Planning Dep., CATARC, Tianjin, 300000, China

Surinder Kumar

Quantum University, Dehradun Highway, Mandawar, Roorkee, Uttarakhand 247167, India

Dr. Seema Tiwari

Associate Professor Applied Science & General Engineering Army Institute of Technology, Pune.

Mr. John Abbas

IBIS Group, Lucknow, UP, India.

Prof. (Dr.) B. K. Sarkar (Patent Guru/ Expert)

MES, Geh Research, India, USA, Singapore, Japan, Austraila, Germany.

Prof. (Dr.) Reena Singh,

MES's, Pillai HOC College of Engineering &Technology, Rasayani Taluka Panvel, Distt. Navi Mumbai, Maharashtra 410207, India

ABSTRACT

This paper studied the roadmap of the hydrogen energy and fuel cell vehicle (FCV) industry in the US, Japan and the EU, and summarized the development goals, promotion paths and development plans of the hydrogen energy industry in developed countries. On this basis, the development prospect of hydrogen energy and FCV industry in China is further analysed based on resource endowment and current situation of FCV industry in China. Finally, by comparing the development plans and characteristics of hydrogen energy and FCV industry in various countries, the deficiencies of hydrogen energy and FCV industry in China are found, and suggestions are put forward for the development of hydrogen energy and FCV industry in China.

INTRODUCTION

With the increase of global vehicle ownership, the accelerated consumption of fossil energy and the aggravation of environmental pollution have become a worldwide problem. For this reason, various countries have conducted a lot of research around the transformation and upgrading of automotive power. Electric vehicles (EVs), hybrid electric vehicles (HEVs), fuel cell vehicles (FCVs), etc. have undergone research and development in the past decades and have become the main direction of automotive power transformation. The reason why hydrogen energy is considered to be the transformation direction of automotive power is that hydrogen energy and electric energy have many similar characteristics such as zero pollution and high efficiency in the process of use. Besides, FCVs have the advantages of long driving range and short refuelling time, which is complementary to EVs [1].

Due to the increasingly prominent role of hydrogen energy in the world energy revolution, many developed countries have issued hydrogen energy roadmaps or action plans [2]. As an important application field of hydrogen energy, the R&D and promotion of FCVs have received strong support from governments of various countries. After years of development, its technical route has gradually matured, and the development of commercialization has continued to accelerate in recent years, which has great reference significance for China. This paper studies the roadmap of hydrogen energy and FCVs in developed countries such as Japan, the US and the EU, analyses the strategies and competitive advantages of various countries, and makes recommendations for the development of China's FCV industry.

HYDROGEN AND FCV ROADMAP IN DEVELOPED COUNTRIES

US

After two oil crises, the US put forward the "energy independence" strategy in the 1970s, and started research on hydrogen energy. In order to realize the vision of hydrogen energy development, the U.S. Department of Energy (DOE) released the *National Hydrogen Energy*

Roadmap in 2002 to encourage companies, research institutions and industry organizations to study the hydrogen energy industry. Since then, the Bush administration have issued important documents to support the development of hydrogen energy. The hydrogen and fuel cell R&D budget reached its peak during the Bush administration. During the Obama and Trump administrations, the US government's enthusiasm for hydrogen energy weakened, so hydrogen and fuel cells R&D budget has been reduced, but still remains at about $150 million a year [3-5].

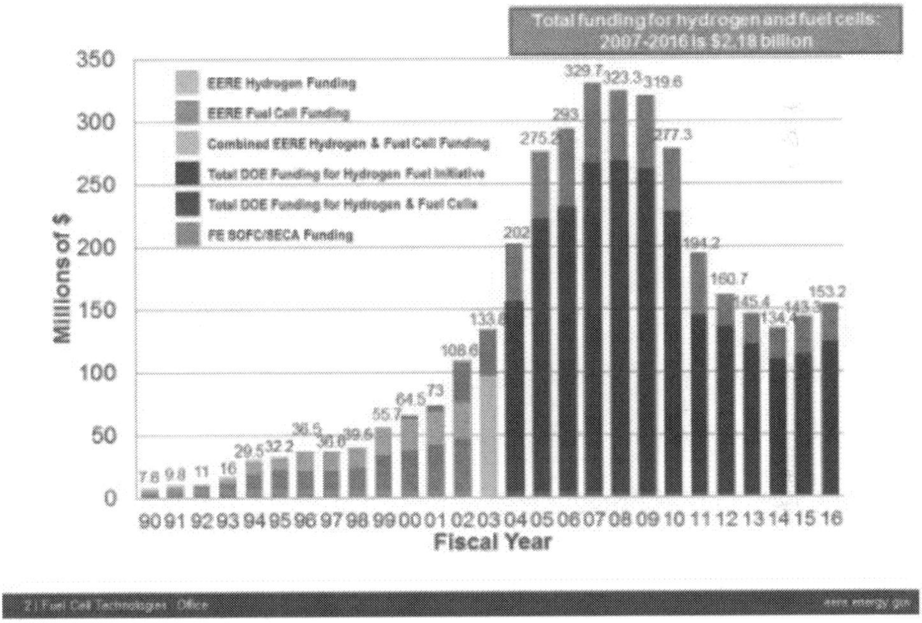

Figure 1: DOE Hydrogen & Fuel Cells Budget History.

Under the long-term high investment of the US government, the US has maintained a world-leading position in hydrogen energy and fuel cell technology, forming a complete industrial chain and a representative enterprise with core technologies [6]. However, due to the inconsistent attitudes of previous governments on hydrogen energy, the continuity of related policies is relatively poor, and the commercialization of hydrogen energy and fuel cell technology is relatively slow [7-8]. Against the background of the acceleration of the global commercialization of hydrogen and FCVs and the maturity of technology, the Fuel Cell and Hydrogen Energy Association (FCHEA) released *Road Map to a US Hydrogen Economy* in 2019, called on the US government to promote the development of the hydrogen economy.

Table 1: Road Map to a US Hydrogen Economy.

	2020-2022	2023-2025	2026-2030	2030 and beyond
Policy	Dependable, technology-neutral decarbonization goals Public incentives Hydrogen codes and safety standards Workforce development programs	Policy incentives to transition from direct support to scalable market-based mechanisms Implementation of decarbonization policy	Transition of policy incentives in fast-following markets from direct support to scalable market-based mechanisms Applications to broaden beyond transport	Reduced/no direct policy support in certain applications when reaching cost parity Robust hydrogen code at federal level
Hydrogen industry	First dedicated hydrogen production for mobility Mid-scale SMR/ATR + CCUS Mid-scale electrolyser plants (10—50 MW)	First large-scale electrolyser plants (50 MW+) First large scale SMR/ATR + CCUS	Development of electrolytic hydrogen production with dedicated renewables and nuclear	Expanding use of hydrogen, enabling further cost reduction and performance improvement
	Development of gaseous and liquid distribution Second-generation FCEVs and refuelling stations for light-duty vehicles, buses, and material-handling vehicles First-generation FCEVs and refuelling stations for heavy-duty vehicles	Hydrogen pipeline/delivery systems in industry clusters New FCEV brought to market Second-generation FCEVs and refuelling stations for HDVs	Development of SMR/ATR + CCS First hydrogen pipelines to connect production sites with demand centres	Retrofitting of reforming capacity with CCUS Competition of electrolytic hydrogen production with SMR/ATR + CCS on cost Variety of vehicle models available
FCV	7600	50,000	200,000	5,300,000
Refuelling stations	63	110	580	5600
Material-handling refuelling stations	120	300	600	1500

he roadmap proposes to promote industrial development through the cooperation of government, industry and scientific research institutions. The government's work mainly includes investment, policy formulation, and legislative supervision. Based on the existing technology, the government should continue to encourage technological research and development to ensure the safety of hydrogen use and improve product performance. At the same time, the development of the hydrogen energy industry requires large-scale hydrogen production facilities, storage and transportation facilities, and refuelling infrastructures. The construction of these facilities requires government investment. In order to increase public acceptance, it is recommended that the US government promote the development of the hydrogen energy industry including policies, regulations, technology, and infrastructure. In terms of policies and regulations, establish direct incentive policies to accelerate the early market commercialization process, improve standards and regulations, and strengthen industry supervision to ensure the healthy development of the industry. At the same time, introduce technology-neutral decarbonization goals to expand the use of hydrogen. With the continuous expansion of the market size, policy incentives transition from direct support to scalable market-based mechanisms. In terms of technology, it is necessary to develop advanced hydrogen production technologies to gradually achieve "zero emissions" in hydrogen production; develop hydrogen energy storage and transportation technologies to reduce the cost of hydrogen use; enterprises and scientific research institutions should develop advanced products and increase public acceptance. Government should increase investment in infrastructure, expand the scale of the industry, reduce the cost of hydrogen production, storage and transportation, and thus reduce the cost of end use. In addition, the construction of refuelling station should be accelerated to prevent infrastructure from becoming a factor hindering industrial development.

Japan

The R&D of FCVs in Japan began in the 1990s. Japanese automobile companies such as Toyota and Honda have successively carried out R&D of FCVs by virtue of rich technological accumulation and the pursuit of innovation. With the successful launch of the Toyota Mirai in 2014, Japanese government accelerated the promotion of the development of the hydrogen energy industry. In 2014, the Ministry of Economy, Trade and Industry of Japan (METI) released *The Strategic Road Map for Hydrogen and Fuel Cells* and *Basic Hydrogen Strategy*, then released *Basic Hydrogen Strategy* in 2017, announcing that Japan will realize a "Hydrogen Society" in 2050 [9-10]. In 2019, METI revised *The Strategic Road Map for Hydrogen and Fuel Cells* to match the *Basic Hydrogen Strategy* and formulated more detailed technical goals.

Table 2: Basic Hydrogen Strategy.

	2017	2020	2030	Future
Hydrogen supply	Hydrogen from fossil fuel		Develop international hydrogen supply chain Develop hydrogen from renewable resources domestically	Zero emission hydrogen production
Hydrogen cost	10 $/kg		Below 3 $/kg	Below 2 $/kg
FCV	2500	40,000	800000	Replace conventional gasoline mobility
FC bus	2	100	1200	
FC forklift	40	500	10000	Introducing large FCVs
Refuelling station	100	160	900	Replace traditional residential energy systems

Table 3: The Strategic Road Map for Hydrogen and Fuel Cells.

		Target	Action plan
Hydrogen supply	Fossil fuel +CCS	Hydrogen cost: ¥30/Nm^3 in 2030	The success of Japan-Australia Brown Coal-to-Hydrogen project. Scaling-up and improving efficiency of brown coal gasifier Enables highly efficient hydrogen liquefaction Low-cost CO_2 capture technologies
	Green hydrogen	Establishment of the technology of hydrogen production from Renewable energy	Expansion of the demonstration in model regions for social deployment utilizing the achievement in the demonstration. Development of electrolyser with high efficiency and durability Development of supply chain utilizing local resources
Hydrogen use	FCV	Achieving a cost reduction of FCV to the level of HV around 2025 Reducing cost of main elemental technologies around 2025	Sharing technical information and problems Reducing the amount of platinum used. Reducing of amount of carbon fiber in hydrogen storage systems
	Hydrogen refuelling station (HRS)	Reduction of cost for construction and operation by 2025 Setting of cost target for each component	Thoroughly integrate promotion of regulatory reform and technological development. Consideration for nation wide networking of HRS Extending opening hours Increasing of the number of HRS with gasoline station/convenience store
	FC bus	Expansion of regions where FC buses run Reducing FC bus's price by half	Developing technology for enhancing the fuel efficiency and durability of such vehicles Expansion of types other than city buses Promotion of deployment of HRS for FC buses
	FC forklift	Expansion to overseas markets	

Constrained by energy resources, Japan faces the problem of high hydrogen use cost. Therefore, the main plans of Japan's hydrogen energy strategy is to obtain cheap hydrogen energy, reduce the cost of hydrogen energy industry, and expand the application of hydrogen energy. In terms of hydrogen energy supply, Japan will establish an overseas supply chain to obtain cheap hydrogen or raw materials for hydrogen production. At the same time, it will implement a "Power-to-Gas" strategy to develop low- cost electrolytic hydrogen production technology and realize domestic low-cost renewable energy hydrogen production in 2032. In order to reduce hydrogen use cost, Japan will develop low-cost hydrogen energy storage and transportation technology to reduce the transportation cost of imported hydrogen. In terms of hydrogen energy use, FCVs are Japan's key promotion direction, and fuel cell passenger cars are the mainstay. Therefore, the Japanese government will accelerate the construction of refuelling station, reduce the purchase and use costs of vehicles, and realize the replacement of traditional fuel vehicles by FCVs in 2050 [11-12].

EU

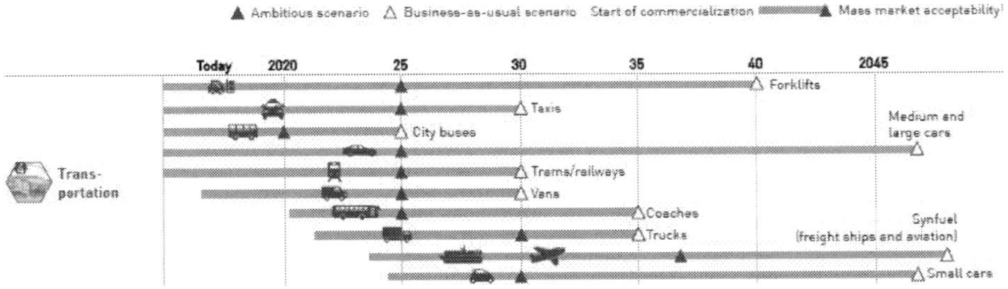

Figure 2: Deployment of hydrogen technology in transportation [13].

In 2019, FCH-JU released the *Hydrogen Roadmap Europe*, proposing that hydrogen energy has huge economic, social and environmental benefits for Europe, and planned the development route of the European hydrogen energy industry in the next 30 years. Based on two scenarios of "ambitious" and "business as usual", the report shows the development prospect of hydrogen and FCVs in Europe [13].

Mass market acceptability is defined as sales >1% within segment as shown in Figure 2. In ambitious scenario, buses are expected to be the first to gain market acceptability in 2020, while heavy trucks and small cars may take 10 years to develop. In business as usual scenario, taxis, trucks, trams, etc. are expected to be accept by the market in 2030, and passenger cars are difficult to promote in a short period of time. With the promotion of FCVs, the number of refuelling stations will also increase. There are currently 120 refuelling stations in Europe. In ambitious scenario, the number of refuelling stations will reach 3750 by 2030, with a total investment of 8.2 billion euros. This number will reach 33,000 by 2050, with a total investment of 27.5 billion euros.

Table 4: Hydrogen Roadmap Europe.

		2030	2050
FCV	Small car	0%	14%
	Larger car	2%	28%
	Taxis	8%	57%
	Vans/LCVs	3%	30%
	Buses	2%	25%
	Trucks	<1%	21%
Refueling station	Amount	3750	33000
	Investment	8.2 billion euro	27.5 billion euro

In terms of hydrogen energy supply, steam methane reforming (SMR)+CCS technology and electrolytic water hydrogen production technology will be the main hydrogen production technologies in Europe in the future. SMR+CCS technology can reduce CO_2 emissions by up to 90%. It has the advantages of low cost and large-scale hydrogen production, and is the main hydrogen production method used in the transition from traditional energy to hydrogen energy. Hydrogen production by electrolyzed water has the advantages of pollution-free and distributed development, and it is expected to significantly reduce costs in the long term. It is suitable for small hydrogen supply sites such as refuelling stations, so it also has development potential.

Europe has many powerful automobile companies, and the R&D of FCV started early. The European Commission and the European Industry and Research Organization launched a public-private joint fuel cell and hydrogen energy action plan (FCH JU) in 2008 to support the development and demonstration of hydrogen and fuel cells in Europe. *Hydrogen Roadmap Europe* pays more attention to the energy transition in Europe and the economic, social and environmental benefits that hydrogen can bring to Europe.

FCV INDUSTRY SITUATION IN CHINA

It is found that the ultimate goal of the hydrogen energy roadmap in developed countries is to promote energy transition, expand the application of hydrogen energy, and finally achieve the leading role of hydrogen energy in the energy structure. Regarding the application of hydrogen energy, FCH JU predicts that in Europe, hydrogen energy consumed in transportation will account for 30% of total hydrogen energy in 2050, and FCHEA estimates that this proportion will reach more than 40% in the US. It can be seen that FCVs will become an important application area for hydrogen energy. Due to the huge differences in natural resources and technologies in different countries, the marketization processes in different countries are also different. This section analyses China's resource endowment and the status quo of the FCV market, and analyses the prospects for the commercialization of FCVs in China.

Resource Endowment

In China's energy production structure in 2017, raw coal accounted for 68.6%, crude oil accounted for 7.6%, natural gas accounted for 5.5%, hydropower, nuclear power and wind power accounted for 18.3% [14]. The energy structure of "rich coal, lean oil, and low gas" has resulted in China's current reliance on coal gasification to produce hydrogen. With the transformation of China's energy consumption structure, the proportion of coal in the energy consumption structure will continue to decline, and the proportion of clean energy will increase [15].

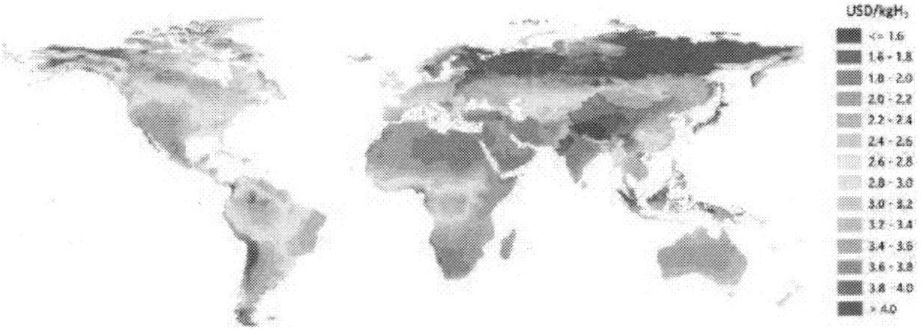

Figure 3: Hydrogen costs from hybrid solar PV and onshore wind systems in the long term [16].

According to the report released by the IEA: China's central and western regions are rich in solar and onshore wind energy resources and have great potential for using renewable energy to produce hydrogen. As the cost of electrolyzed water equipment decreases, the cost of hydrogen from renewable sources in China is expected to fall below 1.6 $/kgH_2$, as shown in Figure 3 [16]. Therefore, hydrogen from coal+ CCS technology can guarantee the recently increased demand for hydrogen in China, and the hydrogen production process is relatively environmentally friendly. In the long run, China's use of renewable energy for electrolytic hydrogen production has more development potential and can accelerate China's transition to clean energy.

FCV Market Situation in China

The development of FCVs started late in China, but as the world's largest automobile market, China has unique advantages in developing FCVs and has made many breakthroughs in demonstration applications. By the end of 2019, China had sold 2,737 FCVs, increased by 79.2% from a year before. A total of 66 refuelling stations have been built, of which 46 have been put into operation [18].

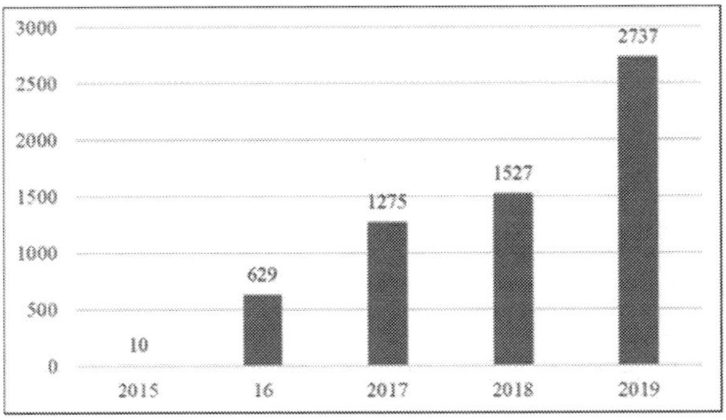

Figure 4: The sale volume of FCV in China.

The commercialization of FCVs in China has been accelerated significantly in recent years, as shown in Figure 4. The important reason is the strong support of the government (Table 5). *2019 Chinese Government Work Report* mentioned promoting the construction of refuelling station for the first time, and *the new energy vehicle development plan (2021-2035) (draft for comment)* mentioned promoting the commercial operation of FCVs and accelerating the construction of refuelling stations. At the same time, a number of local governments have also issued FCV development plans and launched FCV demonstration operation projects.

Table 5: FCV policies in China.

Date	Policy	Content
2009.03	*Auto Industry Adjustment and Revitalization Plan*	Support the development of NEV, including FCV.
2012.06	*Energy Saving and New Energy vehicle Industry Development Plan (2012-2020)*	Catch up with advanced technology in hydrogen and FCV industry.
2016	*Technology Roadmap for Energy Saving and New Energy Vehicles*	FCV development goals from 2015-2025.
2017.05	*Medium- and Long-term Development Plan of Automobile Industry*	Encourage to intensify R&D of FCV.
2019.03	*2019 Government Work Report*	promote the construction of refuelling infrastructures.
2019.12	*Development Plan of New Energy Vehicle Industry (2021-2035) (draft for comments)*	promote the commercialization of FCV, develop refuelling infrastructures, encourage innovation and technology breakthrough.

With the strong support of the government, the scale of FCV market in China has increased significantly. By July 2019, 905 FCVs have been put into demonstration operation in Shanghai, and 768 FCVs have been put into demonstration operation in Foshan. Zhangjiakou has taken the opportunity of the Winter Olympic Games to promote FCVs, and 174 FCVs have been put into demonstration operation [19].

Prospect of FCV in China

The commercialization of FCVs in China is in its infancy. The main driving force for the promotion of FCVs at this stage is government incentives and demonstration operation projects, so commercial vehicles are the mainstay. According to planning goals of *Technology Roadmap for Energy Saving and New Energy Vehicles*, China will start the promotion from fuel cell commercial vehicles, gradually increase the number of fuel cell passenger vehicles, and accelerate the construction of hydrogen supply systems and refuelling stations.

Table 6: FCV development goal in China [17].

		2020	2025	2030
Overall target		Small-scale demonstration and application of 5,000 public service vehicles in specific areas	A large number of applications accumulated to 50,000 in the field of private vehicles and public service vehicles	A total of 1 million private passenger cars and commercial vehicles have been commercialized on a large scale
Hydrogen infrastructure	Hydrogen supply	Renewable energy distributed hydrogen production; co-oven gas and other by-product hydrogen production. Efficient and low-cost hydrogen separation and purification technology.		Renewable energy distributed hydrogen production
	Hydrogen transportation	High pressure gaseous hydrogen storage and transportation	Cryogenic liquid hydrogen transport	Hydrogen storage and transportation of atmospheric high-density organic liquids
	Refueling station	100	350	1000

At present, the cumulative promotion of FCVs in China has exceeded 5,000, which is currently completed in advance by 2020, but the construction of refuelling stations has not yet reached the target. In addition, China should increase its investment in the hydrogen supply system and continuously improve its industrial base in order to achieve the goals of scale in early 2025 and large-scale promotion in 2030.

ENLIGHTENMENT TO HYDROGEN AND FCV INDUSTRY IN CHINA

According to the roadmaps of hydrogen energy and FCVs in various countries, the following conclusions can be made:

Table 7: Comparison of roadmaps.

Country	Characteristic	2030 target	
US	The technology is advanced but the marketization process is slow, so the roadmap focuses on applications such as FCVs	Hydrogen supply	Hydrogen production: hydrogen from NG, nuclear and renewable resources Transportation: focus on pipeline
		FCV	5,300,000
		FCV forklift	300,000
		Refueling station	5,600
Japan	Aiming at a hydrogen society, the emphasis is on FCVs replacing fuel vehicles and FCV exports. Passenger vehicle technology has advantages.	Hydrogen supply	Hydrogen production: import fossil fuel or hydrogen, hydrogen from renewable resources Transportation: Low-cost liquid hydrogen transport
		FCV	800,000
		FCV forklift	10,000
		Refueling station	900
EU	Focus on the social, economic, and environmental benefits of hydrogen, and hope to drive European economic growth. Optimistic about the application of hydrogen in long-distance and heavy-duty vehicles	Hydrogen supply	Hydrogen production: hydrogen from NG, renewable resources
		FCV	One in 12 are commercial vehicles, one in 22 are passenger vehicles, 45,000 trucks and vans
		Refueling station	3750

China	There is a certain gap between technology and developed countries, but the development speed is relatively fast. At present, commercial vehicles are mainly used.	Hydrogen supply	Hydrogen production：hydrogen from coal, renewable resources Transportation: combination of multiple methods
		FCV	1,000,000
		FCV forklift	-
		Refueling station	1000

In contrast, the US, Japan, and the EU have all established the important position of hydrogen energy in the energy system, and FCVs are an important area for hydrogen energy applications. Although China has determined its support for FCVs, it has not paid enough attention to the hydrogen energy industry, which may easily lead to lack of infrastructure. In addition, China's planning for the number of hydrogen refuelling stations is obviously too small. In 2030, the planned number is only 100 more than that of Japan, and China's land area is almost 25 times that of Japan. The number of hydrogen refuelling stations is closely related to the promotion of FCVs. China should increase the number of infrastructure plans and accelerate infrastructure construction. Finally, China's fuel cell core technology still has a large gap compared with developed countries [20], and should follow the establishment of institutions such as FCH-JU and FCHEA, and set up special hydrogen and fuel cell industry R&D institutions to increase technology investment, break through the bottleneck of industrial development and narrow the gap with developed countries.

CONCLUSION

Based on the above analysis, this paper believes that China's hydrogen energy and FCV industry has begun to take shape, but it is still not sustainable. The government should clarify the position of hydrogen energy in China's energy structure and determine the development route and application scope of the hydrogen energy industry. In the future, China should increase investment in technology research and development and infrastructure construction, set up hydrogen and FCV industry R&D institutions to fund the development of advanced technologies, and accelerate the construction of hydrogen energy production, storage, transportation, and refuelling infrastructure construction to promote industrial development.

REFERENCES

Yi B.L. (2018) Development Process and Technical Status of Fuel Cell and Fuel Cell Vehicle. Science Press, Beijing.

Liaowang Institute. (2019) Building China's Strength in Automotive 2019. Xinhua Press, Beijing.

Bian W.Y., Chen T., Chen X.Y., Ge C.L., Hui Z.Y., Yang H. (2019) Study and Enlightenment of Energy Policies of Major Developed Countries. Bulletin of Chinese Academy of Sciences 034(004): 488-496

Du B.G., Zhu R.N. (2018) From Nixon to Trump: Analysis of Evolution of the Policy of "Energy Independence" in the United States for 50 Years. Science and Management 38.208(1): 54- 59+64

CATARC. (2018) Annual Report on the Development of Automotive Hydrogen Industry in China (2018). Social Sciences Academic Press (China), Beijing.

Cao Y. (2019) A Review on Current Situation of Hydrogen Energy Industry in China and the United States. Green Petroleum & Petrochemicals 4: 1-6+19

Shao Z.G., Yi B.L. (2019) Developing Trend and Present Status of Hydrogen Energy and Fuel Cell Development. Bulletin of Chinese Academy of Sciences 34.04:95-103.

You S.J., Zhang Z., Zhou Y. Du G.M., Wan H., Xu P., Bao Q.L. (2019) Industrial Policies of Advanced Hydrogen Energy Countries and Enlightenment. Oil Forum 38.05:57-66

Zhen Z.J. (2016) Analysis of Industrial Technology & Strategy Roadmap of FCV in Japan.

Advanced Technology of Electrical Engineering and Energy 35(7): 50-54

Wang J., Zhou W., Yao Z.H., Fang H.F. (2019) Research on the Development of Experience and Enlightenment of Fuel Cell Vehicle Industry in Japan. China Auto (12): 36-39

Ministerial Council on Renewable Energy, Hydrogen and Related Issues. (2017) Basic Hydrogen Strategy. https://www.meti.go.jp/english/press/2017/pdf/1226_003b.pdf

Hydrogen and Fuel Cell Strategy Council. (2019) The Strategic Road Map for Hydrogen and Fuel Cells. https://www.meti.go.jp/english/press/2019/pdf/0312_002b.pdf

FCH-JU. (2019) Hydrogen Roadmap Europe: A Sustainable Pathway for the European Energy Transition. https://www.fch.europa.eu/publications/hydrogen-roadmap-europe-sustainable-pathway-european-energy-transition

Liu P.K., Wang Z.W. (2019) Is It Reasonable for China to Promote "Energy Transition" Now?

——An Empirical Study on the Substitution-Complementation Relationship among Energy Resources. China Soft Science 8:14-30

Xie H.P., Wu L.X., Zheng D.Z. (2019) Prediction on the energy consumption and coal demand of China in 2025. Journal of China Coal Society 44(7):1949-1960

IEA. (2019) The Future of Hydrogen. https://www.iea.org/reports/the-future-of-hydrogen

SAE-CHINA. (2016) Technology Roadmap for Energy Saving and New Energy Vehicles. China Machine Press, Beijing.

China Hydrogen Alliance. 2020.01 Hydrogen Report. http://www.h2cn.org/dynamics_detail/643.html

CATARC, Royal Dutch Shell Group of Companies. (2019) Annual Report on the Development of Automotive Hydrogen Industry in China (2019). Social Sciences Academic Press (China), Beijing.

Liu Z.W., Shi T.Z., Hao H., Zhao F.Q. (2018) Research on Main Problems Associated with Development of Fuel Cell Vehicle in China. Automotive Technology. 1:1-9

Made in the USA
Monee, IL
12 November 2026

f7e416ce-1240-4654-b351-1e874a3b02ecR01